DI064347

Advocacy
in
Health Care

Advocacy in Health Care

The Power of a Silent Constituency

Edited by

Joan H. Marks

Humana Press • Clifton, New Jersey

Library of Congress Cataloging-in-Publication Data

Advocacy in health care.

 (Contemporary issues in biomedicine, ethics, and
society)
 Proceedings of the Advocacy in Health Care Conference,
held at Sarah Lawrence College in Mar. 1984; sponsored
by the Health Advocacy Program of Sarah Lawrence
College.

 Includes index.
 1. Hospitals--Patient representative services--
Congresses. 2. Patient representatives--Congresses.
3. Medical care--Quality control--Congresses. I. Marks,
Joan H. II. Advocacy in health Care Conference (1984 :
Sarah Lawrence College) III. Sarah Lawrence College.
Health Advocacy Program. IV. Series. [DNLM:

1. Patient Advocacy--congresses. W 85 A244 1984]
RA965.6.A35 1986   362.1'042   86-283
ISBN 0-89603-092-X

© 1986 The Humana Press Inc.
Crescent Manor
PO Box 2148
Clifton, NJ 07015

Printed in the United States of America

Acknowledgments

The Advocacy in Health Care Conference was supported, in large part, by a generous gift from Pfizer, Inc. Their assistance was not only financial, but extended to substantive discussion of the need to interest the broadest possible health-care constituency to consider the issue of advocacy in health care.

Many people were involved in conference planning and preparation of this volume. Doctor Richard Nesson supported the concept of the conference and served ably as Conference Chairman. Ruth Ravich and Elinor Miller were central to the program selection; Beatrice Greenbaum served as Conference Coordinator and was ably assisted by Molly Colosimo. The support staff at Sarah Lawrence College, skillfully organized by Diane Brewer, facilitated the complex arrangements necessary for the Conference. Helene Friedman provided invaluable editorial expertise in the preparation of this volume.

Preface

The roles of both the consumer and the health advocate professional have become increasingly significant in today's climate of "rationed" health care. It seems clear that the timely exchange of ideas among seasoned health care advocates is necessary if we are to deal with the complex problems of a technologically advanced society seeking to ration its heath care in a truly humane way.

Toward such a timely exchange, the first Conference on Advocacy in Health Care was organized by the Health Advocacy Program of Sarah Lawrence College and recently held. *Advocacy in Health Care: The Power of a Silent Constituency* is the proceedings of the conference and will, we believe, greatly extend our efforts to share both the problems and solutions that effective patient advocacy entails. Never before has the issue of advocating for special population groups by combining the resources of consumers and professionals been the exclusive focus of one volume. This book discusses the power of such an alignment and describes specific organizational techniques that have been effective in bringing about changes in the delivery system.

The final section of the book, "Questions, Comments and Answers," presents a selection of topics of special interest that surfaced during the open discussion at the last conference session. The comments were forthright in their criticism of public policy, and the vigor of the argument underscored the vitality of the coalition between professionals and consumers.

The Health Advocacy Program, the first graduate program of its kind, was developed in 1980 for the purpose of training patient advocates to serve as ombudsmen in hospitals, health-care centers, and long-term-care facilities. The primary goal of the conference was to draw attention to the accomplishments achieved by health-care advocates, both professional and lay, and to foster an exchange of ideas among people working in the many different constituent areas of health care.

Joan H. Marks

Introduction

Advocacy in Health Care reports the proceedings of an unusually interdisciplinary conference held at Sarah Lawrence College, Bronxville, New York. The more formally presented and augmented articles presented here describe the development of a powerful consumer health advocacy movement concerned with health-related services and institutions throughout the United States. This movement is considered by some to be conceptually rooted in the government ombudsmen system mandated by Danish law in 1809, and the present American movement is perceived to be a general outgrowth of our frustration with the rapidly escalating costs of health services, the difficulties in accessing necessary and appropriate services, and the frequency with which health service users have suffered alienating experiences at the hands of insensitive providers. Both health-care professionals and consumers recognize the need to change this unsatisfactory situation. Five hundred health-care consumers and professionals came together to address shared concerns, to identify the system's problems and deficiencies, and to explore ideas about changing health-care practices. These concerns, problems, and ideas are each treated in some depth in this volume.

Governmentally, America has not yet established an individual's right to health, but as a nation we speak about one's right to health care. The President's Commission for the Study of Ethical Problems in Medicine and Biomedical and Behavioral Research stated that a decent level of medical care is the right of all citizens. At the same time, they noted that, in 1977, 34 million Americans were without health insurance. If priority is to be assigned to the concept of the right to medical care, then public policy decisions must be directed toward the reallocation of national resources to meet this end. Presently, our national ethic appears to emphasize achieving the benefits of a wealthy, technologically advanced society. It is a widely, but by no means universally, held view that this should include a commitment to meet the health service needs of all Americans in an effective and compassionate fashion. Disease prevention and health maintenance, as well as health care, must be part of this commitment.

To address the shared concerns of health-care providers and consumers who recognize the present system of delivering care as inadequate, the *Health Advocacy Program* at Sarah Lawrence College invited a distinguished group of health advocates to present their views and experiences at this conference. These participants considered the causes of our current deficiencies and shared their ideas for and approaches to improving the system. We believe that the conference and these published proceedings will further motivate health advocates to work for needed changes.

Reviewing the milestones achieved by public-interest advocacy groups, the Honorable Paul G. Rogers refers in his paper to the legislation protecting mental patients, prisoners, the poor, and the elderly, and notes that the enforcement of these hard-won laws has gone beyond the capacity of public-interest groups to assure. The need for individual advocates to help patients assert their health-care rights has thus become even more essential.

Mister Rogers sees an additional important role for patient advocates—to assist patients to understand the choices they have the *right* to make concerning treatment. He points out that helping patients understand about their health care is not only the fair, but also the therapeutic, thing to do. Mister Rogers believes advocates can serve an important function by supporting patients in resisting unwanted treatment—a right he feels should be effectively upheld. Noting that hospitals are already being rewarded for providing fewer therapies, he reminds the audience that patient advocates may increasingly be called upon to protect the patient's right to receive all appropriate and available medical care.

Commissioner Axelrod sees a great need for health advocates and asks why those who stand to benefit most from medical care are often the ones least likely to gain access to care. He sees private, entrepreneurial medical practice as the promoter of social problems stimulating these regulatory mechanisms. Doctor Axelrod, in his article, views the difficulties that arise from the cost, accessibility, and quality of health care to be a result of medical practitioners' abuse of the social contract, rather than as a consequence of too much regulation. Quoting Milton Roemer, Dr. Axelrod reminds us, "the social physician can no longer be regarded as an ethereal ideal of scholars and dreamers; he or she has come to be expected by people everywhere."

Robert Butler, addressing the question of advocacy for the elderly, refers to the triage system of care practiced today and says, "We are still in a battle, but it's a battle of the budget." Although there has been a 27-year gain in the average life expectancy since 1900, it is not yet a cause for celebration because we have not adjusted to this "longevity revolution." Advocates must organize to prevent the continued blaming of the elderly for these problems, and to work for reforms in reimbursement that will allow comprehensive diagnosis, care, and management of patients, regardless of age. Doctor Butler notes the failure of schools of medicine, nursing, and social work to teach either current concepts of human aging or the related social and clinical problems. Minimal research funds have been directed toward solving these problems, and this trend must be vigorously contested.

Myrna Lewis focuses on elderly women and points out that, since 6½ million more women than men voted in 1980, women should rightfully be considered politically powerful. She believes their recognized interest in health-care issues can, and should, be directed toward advocating better services for the elderly.

In addressing the issue of advocacy for children, Sara Rosenbaum classified America's health policy toward children as nothing less than a "national tragedy." The range of strategies available for child advocates includes legislative advocacy, litigation, public education, and research. Since 1979 there has been a 31% increase in the poverty rate of children, resulting in 13 million children whose needs must be addressed by child-health advocates. Noting the need to assume a broad perspective toward medical care, Sara Rosenbaum encourages a community-based approach to increase county, city, and state expenditures for poor mothers and children. At the Congressional level, basic programs of health care must first be maintained, and then increased. Newer technologies make sense when they are coordinated with community-based followup and outreach.

George Lythcott challenges those involved in advocacy for minority groups to achieve the goal of reducing the morbidity caused by preventable disease states. This involves the effective dissemination of health information, as well as attitudinal changes among minority groups, even when they have an impact on traditional beliefs. Doctor Lythcott reminds us that when we speak of minorities we are referring to approximately 14.6 million Hispanics, 1.4 million American Indians, 3.5 million individuals of Asian/Pacific island ancestry, and 26.5 million blacks. The diffusion of health information is frustrated by the cultural diversity of these groups—a fact not always adequately appreciated by health planners. Doctor Lythcott cautions us to use acceptable international criteria when comparing the health needs of different minority groups, as well as to appreciate the respective roles both the larger and core communities must play.

The hospice movement described by David Greer is the most successful advocacy effort in the United States in the last decade. This quintessential advocacy venture mobilized a powerful constituency and has further led to a new national system of health care in which nonhospital care has become reimbursable. Doctor Greer reminds us that dying has become accepted as an important part of living, and our health delivery system is now expected to provide for an improved quality of life that includes the period of death.

Doctor Greer believes group advocacy can be organized by professionals, clients, and/or individuals around any specific concern. Each constituency has special powers and when acting in concert can be uniquely successful. In the case of the hospice movement, all parties have advocated humane treatment within the established network of the family. Hospice advocates, particularly volunteers, have organized professionals from various disciplines and coordinated lobbying activities that eventually led to passage of legislation making hospice care reimbursable—ultimate proof of the success of the hospice movement.

Irving Selikoff addresses the issue of health on the job. His paper, titled "Confessions of a Naive Scientist," reveals his belief that obtaining scientific facts is not enough, nor is it adequate to understand what causes disease. Knowledge can lead to the prevention of disease if that scientific information is translated by advocates to make it accessible and usable by consumers and legislators as they deal with environmental issues. For Dr. Selikoff, "science is necessary but not sufficient."

Victor Gotbaum focuses on the hierarchical nature of a health-care system that fails to recognize all workers as essential components of the system. He reminds us that there are still major parts of the hospital work force that are not viewed as key contributors to the patient's care. Recognition of these workers is necessary before they will achieve "part of the action." In organizing these paraprofessionals effectively, one must carefully define the workplace and its membership and acknowledge that success is possible only when a thorough analysis is effectively presented. Finally, Mr. Gotbaum challenges the group to focus on the patient, to think of the entire staff as the team, and to recognize that everyone has an important role to play in administering patient care.

Advocacy for the mentally ill has had considerable success both in reducing the stigma of mental illness and in developing mental health programs throughout the country. But the mentally ill themselves are not the best advocates, except in the rare instances in which they are well-known public figures. Herbert Pardes points out that recognizing the suffering caused by mental illness is not enough to motivate legislators to provide adequate funds for care of the mentally ill or for research into the causes of mental disorders. Professional and citizen groups can complement each other by getting across the message that society will achieve better mental, social, and economic health if we invest now in mental health programs. Research efforts in mental disease have been remarkably successful in developing both a psychopharmacological and psychosocial approach to caring for the mentally ill, and the efficacy of these interventions must be communicated to decision makers.

Don Harper Mills approaches the topic of malpractice in health care from the point of view of the provider. He sees the prospective payment system as the cause of an economic crisis for physicians and hospitals. Doctor Mills believes it is not economically feasible to switch from a fault system to a nonfault system if the source of funding remains with members of the health care team. Changing the system of funding will require tax support or a combination of hospitalization, major medical, and disability insurance with medical adverse-outcome insurance. He notes that improved care could result from an improved peer review system, as well as more effective quality assurance programs.

Vergil Slee sees the area of quality assurance as one of quality management. The key question is, "Will the price for lower cost be lower quality?" He believes that this is a real danger. Observing that advanced (higher cost) technology is not equivalent to better-quality care, Dr. Slee points out that medical care no longer involves mainly the physician and the patient, but now includes many other health-care workers who assist in carrying out medical procedures. Any failure in this chain of providers can adversely affect the quality of the total care. Quality management thus relates directly to individual performances that we are in fact able to evaluate. The challenge is to manage these performances effectively.

Ruth Ravich discusses the field of patient advocacy or patient representation. She describes the new health professional in this field as a "generalist with a broad overview of the institution and its formal and informal rule system." By seeking out patients needing assistance, the patient advocate can focus on problem areas before problems become unresolvable. As a patient educator, the advocate facilitates communication between patients, families, and health professionals. All these functions serve to protect patients' rights in a way that is beneficial to both patients and institutions. Experience has shown that skilled patient representatives or advocates can be on the administrative staff of a health-care facility without compromising their ability to meet patients' needs and rights.

Suzanne Rauffenbart addresses the issue of communication, noting that there is frequently no systematic approach to communication with patients. She discusses the methods used to change patient attitudes, but observes that much information offered patients in hospitals is designed to elicit what management needs to know without sufficiently recognizing what the patient's information needs may be.

The last section of the book, Questions, Comments, and Answers, presents a selection of the dialog in the open discussion period at the final conference session. Several topics of special interest to the conferees are highlighted. The comments were forthright in their criticism of public policy, but the vigor of the argument underscored the vitality of the coalition between professsionals and consumers.

Contributors

David Axelrod, MD, Commissioner, New York State
Department of Health, Albany, New York

Robert N. Butler, MD, Brookdale Professor of Geriatrics
and Adult Development, Chairman, Gerald and
May Ellen Ritter Department of Geriatrics and
Adult Development, Mt. Sinai School of Medicine,
New York City, New York

Victor Gotbaum, MIA, Executive Director, District
Council 37 New York City, New York, American
Federation of State, County and Municipal Em-
ployees AFL-CIO

David Greer, MD, Dean of Medicine, Brown University
School of Medicine, Providence, Rhode Island

Myrna Lewis, ACSW, Department of Community Medicine, Mt. Sinai School of Medicine, New York City,
New York

Joan H. Marks, Codirector, Health Advocacy Program,
Director, Human Genetics Program, Sarah
Lawrence College, Bronxville, New York

Paul G. Rogers, JD, Hogan & Hartson, Chairman, National Council on Patient Information and Education, Washington, DC

George I. Lythcott, MD, Professor of Pediatrics,
Edward Jenner Professor of International Health,

University of Wisconsin-Madison, School of Medicine, Madison, Wisconsin

Don Harper Mills, JD, Clinical Professor of Pathology, University of Southern California School of Medicine, Los Angeles, California

Herbert Pardes, MD, Director, New York State Psychiatric Institute, New York City, New York

Suzanne Rauffenbart, Vice President, Public Affairs, Memorial Sloan-Kettering Cancer Center, New York City, New York

Ruth Ravich, Director, Patient Representative Department, Mt. Sinai Hospital, New York City, New York, Codirector, Health Advocacy Program, Sarah Lawrence College, Bronxville, New York

Sara Rosenbaum, JD, Director, Child Health Children's Defense Fund, Washington, DC

Irving J. Selikoff, MD, Director, Environmental Sciences Laboratory, Professor of Community Medicine, Mt. Sinai School of Medicine, New York City, New York

Vergil N. Slee, MD, President, Slee & Associates, Brevard, North Carolina

CONTENTS

Preface . . . . **vii**
Introduction . . . . **ix**

Milestones in Public Interest Advocacy **1**
Paul G. Rogers

The Need for Health Advocacy **9**
David Axelrod

Hospice as Advocacy **19**
David Greer

Health Advocacy Among Minority Groups **27**
George I. Lythcott

Advocacy for Children **41**
Sara Rosenbaum

Patient Advocacy **51**
Ruth Ravich

Advocacy for the Elderly **61**
Robert N. Butler

Advocacy Issues for Older Women 67
Myrna Lewis

Unionized Health Workers .77
Victor Gotbaum

Quality Management neé Quality Assurance 83
Vergil N. Slee

Confessions of a Naive Scientist 95
Irving J. Selikoff

New Approaches to the Resolution of Profes-
    sional Liability Problems 99
Don Harper Mills

Advocacy for the Mentally Ill 103
Herbert Pardes

Communication Networks: Advocacy in Health
Care 113
Suzanne Rauffenbart

Comments, Questions, and Answers 119

Biographies 129

Index 133

Milestones in Public Interest Advocacy

Paul G. Rogers

Introduction

Advocacy in health care is a calling many of us have pursued—one way or another—for many years. And yet, it has not attained the full status of an independent profession. I hope—and anticipate—that the outstanding program at Sarah Lawrence and proceedings such as this one will help to accomplish this important goal.

I would like to observe that we do not hear much in Washington these days about protecting the vulnerable. Instead, we hear that there is no hunger in America, and that those who sleep on the streets do so by choice. Yet, some of these street people are patients released from mental hospitals who now live as outcasts not only of society, but also of our health and welfare system.

When I was in Congress, we enacted laws to provide essential health care to those who would otherwise have to do without. Today, those same people are being asked to pay more coinsurance and deductibles than they can possibly afford. I am well aware of the need to contain the costs of heath care, but we must find ways to do this without denying health care to those who are already overburdened. As individual patient advocates, or as advocates for groups of disadvantaged people this will be a constant challenge.

1

Just over a hundred years ago, the American Society for the Prevention of Cruelty to Animals went to court to act on behalf of a child—and was permitted to do so on the grounds that a child is an animal. A society for the prevention of cruelty to children was not established until 1875—ten years later. We have come a long way since then, but we have much more to do.

Public Interest Advocacy

As I reviewed the description of the health advocacy program at Sarah Lawrence, I recalled milestones achieved in the last few decades by public interest advocacy groups and coalitions. In the 1960s and 1970s, they pushed for legal protection for mental patients, prisoners, children, the poor, and the elderly. The implementation and enforcement of the laws that emerged, however, is now beyond the capacity of public interest groups to ensure. That is why it is important that we have individual advocates—within the various programs—to assist patients in knowing and asserting their rights.

In 1973, the Health, Education and Welfare (HEW) secretary's Commission on Medical Malpractice suggested that patient grievance mechanisms would be an effective way to reduce the number of malpractice suits. The idea was that if patients' complaints could be dealt with at an early stage, disagreements could be straightened out, conflicts could be resolved, and litigation could be avoided. The Malpractice Commission had high hopes for the mechanism. They pointed out that the Health Services and Mental Health Administration was already supporting five demonstration projects involving patient advocates in nursing homes. But since then, the funding agency has been reorganized and renamed—HEW has become the Department of Health and Human Services (HHS), and the patient advocacy idea appears to have lost priority in the shuffle. I know of no federal regulation, for example (except in the area of research with human subjects), that requires the presence of a patient advocate in a health-care facility.

Government Regulations and Advocacy

Ten years ago, Congress created the National Commission for the Protection of Human Subjects of Biomedical and Behavioral Research. That Commission was asked, among other things, how to protect children, mental patients, and prisoners who are asked to become subjects of research. Many of the Commission's suggestions have now been incorporated into regulations governing all research supported by federal funds. (New York state has enacted similar provisions.) Those suggestions include the requirement that a representative of the subject population sit on the local review board that decides whether or not proposed research is ethically acceptable. In addition, for certain types of research, the Commission recommended that an advocate be appointed to assist the children, prisoners, or mental patients to understand the proposed research and to decide, free from duress or fear of adverse consequences, whether or not they really wanted to participate.

The Commission recognized that other research subjects, particularly the elderly, might need similar protection because they are often alone, without family members to help them, and could be totally dependent upon an institution for their daily care. For these same reasons, elderly patients in hospitals and nursing homes are likely to need and should be assured of having the assistance of a patient advocate.

The idea of patient advocates came from the patients' rights movement in the 1960s and early

1970s—an outgrowth of the consumer and civil rights movements and the general distrust of government and the "establishment." Conferences were held, articles were published, and activists spread the word in the lay press.

Public Awareness

We have a rule of thumb in Washington that an idea has reached true public awareness when there are jokes about it on the Johnny Carson show. Well, by that standard, patients' rights had reached the collective consciousness by January, 1973, when Johnny Carson read a parody of a patients' rights statement. His list included the following provisions:

- No patient shall be denied the right to seek further medical consultation if he is given an autopsy.
- A patient has a right to assume that if he is in a coma, he will not be used as a doorjamb.

Rather sick, but clearly patients' rights were viewed as so self-evident that they shouldn't have to be written out.

The rights are not self-evident, however, and they are not self-enforcing. This plain truth was pointed out quite forcefully in a law review article that recommended special training for patient advocates who could be placed in hospitals to assure that patients know, and have assistance in asserting, their rights. The authors noted that fewer than half of the 1000 hospitals responding to a 1974 survey had an employee called a patient representative. And those that had such an employee thought the employee's role was to respond to complaints about housekeeping problems. Patient representatives, in those days, had neither the training nor the authority to deal with concerns about health care. That was a decade ago, and the need persists.

Recently, a local TV station in Washington aired an editorial suggesting that health-care facilities be required to read a list of rights to patients at the time of admission—similar to a health-care version of the "Miranda" warning that police are required to give before questioning someone in their custody. The editorial pointed out that although some hospitals have patient advocates who are supposed to protect patients' rights, the advocates are employees of the hospital. Therefore, they could be expected to protect the facility at all costs.

All of us recognized the conflict of interest that may arise when an advocate is hired and paid by the health-care facility. Can he or she be fired for doing too good a job? How can the advocate's freedom to act on behalf of the patients be protected against retaliation (or restriction) by the hospital? This concern must be carefully watched and patient advocates must be alert to withstand such pressures.

Patient Consent

I think that perhaps one of the most important roles for a patient advocate is to help patients understand their condition, and the choices they have to make concerning treatment. Too many people think that "informed consent" is the same thing as a signature on a consent form. But the signature is meaningless if no information has been conveyed. Although studies show that the average American has about an eighth grade education, most consent forms are as difficult to read as articles in scientific journals. They are often terribly complicated documents. (I'm afraid the legal profession must take most of the blame for this.) Often the very young, and very old, the very sick, and those who speak another language have great difficulty understanding the consent form. Unless someone is available to explain the form in terms they can grasp, their so-called consent is a sham.

We also know that the more patients understand about their condition and their treatement, the better they respond—and the quicker they recover. So, helping patients to understand their health care is not only the fair thing to do, it is also therapeutic.

Death With Dignity

Of course, not all people who are sick get well. Some must adjust to the fact of their dying, and the families of those patients must be helped, too. Patient advocates, properly trained, can be of great service to dying patients and their families.

We have all heard a great deal recently about death with dignity and the right to refuse treatment. It would seem from the highly publicized cases that people need help in resisting unwanted treatment. Few would argue, today, about the right of a competent adult to refuse treatment—even lifesaving treatment, except in rare cases. But when a patient becomes incapable of making choices, or of expressing preferences, then help may be necessary.

On the other hand, there is reason to believe that help of a different sort may be needed. Just as patients have a right to resist treatment, they also have a right to resist death and dying. New payment procedures, designed to reduce the costs of health care, will be reversing the incentives for hospitals and other providers. Although they once made more money by increasing the number and complexity of interventions, they soon will make money by limiting care as much as feasible. Hospitals will be rewarded for providing fewer therapies and for discharging patients sooner, rather than later. What will happen to the terminally ill, or the very old, who want to continue aggressive therapy and fight to the end? The patient advocates of the future may well be called upon to protect such patients' rights.

Of course, there will have to be limits to the care a patient (or a patient's family) may insist upon. At some point, the provider will have to be able to make a judgment that certain treatments will provide no medical benefit to a particular patient. But when such decisions must be made, we must be sure that someone will represent the patient's interests in the decision-making process. That person may well be a patient advocate.

The Need for Health Advocacy

David Axelrod

Introduction

As Commission of Health for the State of New York, I am responsible for health advocacy for all the residents of the state. One of the primary reasons the Commissioner of Health is by statute a physician is that he or she is expected to be a forceful advocate of patients' rights, particularly in our health-care institutions. There are specific bills of rights for nursing homes and hospitals, but unfortunately patients often know little about them. For this reason alone, the role of patient representatives in hospitals and in nursing homes is a very important one. They should be viewed as personal proponents and defenders of patients' rights and dignity, with responsibilities to patients that transcend the traditional employer–employee relationship.

Patient Advocacy and Hospitals

This is a particularly appropriate time to discuss this topic because I believe that there is a very insidious threat to the whole process of patient advocacy. This threat is embodied in the increasing number of investor-owned proprietary hospitals, from which we in New York are generally shielded. Those of you who

9

are not aware should realize that in New York state no publicly owned stock company can own a health-care institution unless it is a physician-corporation incorporated within the state of New York, or a partnership of individuals who live in the state of New York. All other health-care institutions are voluntary or public in nature. This may appear to be an extraordinary constraint upon the operation of health-care facilities, but there is a purpose—and the purpose is patient advocacy. The purpose is to ensure that a designated person or persons for each institution can be held accountable. There is a Board of Directors or Trustees for every facility, and they are responsible for the quality of care provided within that institution. One need not go to the pages of the *Wall Street Journal* to discover what company owns the institution that is providing the health care. One can go directly to the Board of Directors of the institution and find out who is responsible for the kind of care being rendered.

I believe there is a clear need to induce more socially responsible and accountable behavior by health-care providers. This would not be necessary if a socially oriented code of medical ethics were followed, and if medical education—for all health-care providers—were recast in a way that would clarify the community health initiatives and policies required to meet the social responsibilities that are incumbent upon each and every one of us. Such clarification and change does not seem imminent.

There are a number of problems that will influence the role of patient advocates. The foremost problem is economic in nature and relates to the burgeoning supply of physicians and the demands that they are placing upon our health-care resources—hospital beds, equipments, dollars. This matter ultimately leads to the question of how much health care we can afford, how to cover costs of caring for the chronically ill, and who shall live and who shall die. I think this is the quintessential issue facing those who serve as patient advocates.

Health Care Etiquette

Uwe Reinhardt, the medical economist, addresses these concerns in a rather witty and learned paper entitled "Table Manners at the Health Care Feast." He observes that, "There may soon come a time when guests presenting themselves at the health-care feast will outnumber the seats at the table and those lucky enough to be seated will scramble for the offerings." He goes on to say that eventually, "table manners may deteriorate both audibly and visibly and society will be forced to meet the issue of manners at the health-care table head on. The question then arises who is to enforce the manners at the table and what book of etiquette is to be used as a guideline." Reinhardt ruefully concludes that a good book of etiquette for health care should take into account three considerations.

"First, it should allow guests free access to the dinner table and not permit a few strong ruffians to chase away daintier guests when the offerings at the table are scarce. Second, it should force the guests at the table to treat the bearers of fiscal nourishment, the patients, courteously and with good care. And third, the various guests should observe some propriety in determining the size of portions they scoop for themselves at the table. Heavy eaters who seek to get their arms up to their elbows in the salad bowl somehow ought to be sanctioned."

Those sanctions can come from advocacy. It is going to be the advocates' responsibility to ensure that the allocations are appropriate and that everyone at the table receives adequate nourishment—the kind of healthful nourishment and care that we have a responsibility to provide.

Three centuries ago Gilles Menage defined medicine as the art of keeping a patient quiet with frivolous reasons for his illness and amusing him with remedies good or bad until Nature kills him or cures him. In this century, Lawrence Henderson, the late, famed biochemist, observed that, "Somewhere between 1910 and

1912 in this country . . . a random patient with a random disease, consulting a doctor chosen at random had, for the first time in the history of mankind, a better than 50–50 chance of profiting from the encounter." In citing Henderson's premise in a 1977 editorial in the *New England Journal of Medicine,* Dr. Franz Ingelfinger, then editor of the journal, commented, "Henderson's words imply progress after 1912, but . . . our astute insurance companies, whose business it is to take no undue risks, apparently have found no actuarial reasons that would persuade them to impose higher premiums on those who, for either religious or philosophic reasons, reject the custody of medicine." Doctor Ingelfinger, who was a distinguished gastroenterologist, went on to say that if we assume that 80 percent of patients have either self-limited disorders or conditions not improvable even by modern medicine, then the physician's actions will not affect the basic course of such conditions. Ingelfinger claims, citing examples, that medical intervention is dramatically successful in slightly over 10 percent of cases. "But, alas, in the final ninety percent, give or take a point or two, the doctor may diagnose or treat inadequately or may just have bad luck. So the balance of accounts ends up marginally on the positive side of zero."

Those seeking to provide advocacy must bear in mind that although the physician is billed as a healer, he or she is not the healer, but a facilitator. The patient, not the physician, heals himself or herself. I think that there has always been an assumption that physicians have great powers. This mystique reaches back to biblical times, when groups of people were thought to have mysterious powers, and it is these powers that were transmitted to physicians. But the physician has none of these strange powers . . . he or she is a human being, like you and me.

Health-Care Costs and
Physician Overabundance

Although one may lend more credence to medical efficacy than that accorded by Menage, Henderson, or Ingelfinger, the role of medicine in influencing human health is of particular concern today as our nation experiences a physician boom that threatens to inundate us with doctors and drive medical-care costs beyond their already unacceptably high levels. Between 1978 and 1990, the nation's supply of physicians is projected to grow from 375,000 to 540,000, a 43 percent increase. By the end of the century, based on current output levels, America will have about 650,000 active physicians, one for every 400 citizens. In New York, according to a study recently issued by the State Education Department, there are some 50,000 active physicians, or one for every 350 of the state's citizens. In the study providing these data, the authors soberly concluded that the current aggregate supply of physicians in New York exceeds the estimated number needed.

Our concern about the oversupply of physicians is tied to the fact that physicians order or authorize between 70 and 80 percent of all health-care spending—health-care spending that in some cases may not be necessary, health-care spending that we need to evaluate. The nation's health-care bill exceeds a billion dollars a day, with almost half of this money provided by federal, state, and local governments. The increased supply of doctors in one of the reasons that health-care costs continue to rise at a far faster rate than the rest of the economic indicators. Unless we address this issue, we will hear a great many more discussions of rationing health-care for the patients whose interest all advocates represent.

Rationing Health Care

Before discussing rationing health care, we must first talk about the rationalizing of our health-care system to ensure that the resources will be there for the patients. Members of Congress and state legislators apparently subscribe to what political scientist Aaron Wildavsky has dubbed "The Great Equation," namely that medical care equals health. They equate huge sums of money spent for medical care with the seldom fulfilled belief that public health gains will result for individual patients. However, as most public health experts acknowledge, and our elected leaders must come to understand, many of our most challenging health problems—for example, the morbidity associated with adolescent pregnancy, smoking-related diseases such as lung cancer, accidental injury, hunger and nutritional deficiencies, and environmental illness—are usually beyond the influence of medical intervention.

Money would be better spent preventing these diseases rather than treating them after they occur. As laudable and deserving of passage as many of the proposals for better health prevention are, these are but a precursor to more far-reaching action that must be taken. Congress should do more to launch a national program to determine how many doctors this nation needs, to establish a formula to reduce the number of medical schools and their enrollments, and to stem the influx of graduates from the so-called offshore diploma mills.

There is a need as well to develop a redefinition of the role of the physician as a health-care provider. Those who are providers of health care must be both scientists and social workers. They must be prepared to cooperate, to use team work, and to remain in close touch with the people they serve—a factor that I think we have all missed. We need the advocate's help in the development of a social physician who protects the health of the people and guides them toward a healthier and happier life; the kind of physician who puts the patient's welfare above pecuniary gain or any other doctrine that may interfere with social responsibility.

Our nation has never had a greater need for health advocates. Despite our huge and unmatched investment in health care, those most in need of medical care—the pregnant adolescent, the poor and downtrodden, and the infirm elderly—are those most commonly neglected by our system of medical and health care. They are the ones least likely to gain access to appropriate, high-quality medical care. Why is it that those who stand to benefit most from medical care often have the greatest trouble securing it? I suspect the answer is that providers of care continue to dominate entry into the system. If you are poor and poorly educated, you do not choose the kind of care that you get, you get the kind of care that the system begrudgingly supplies. All too commonly that means you do not end up in a private doctor's office; you find yourself instead in the coldly impersonal setting of a crowded emergency room or outpatient clinic, waiting to see a doctor of unknown quality.

According to Lois Pratt, in her paper on the doctor–patient relationship. "Persons who obtain the best medical care, both preventive and restorative, are those who strive for mastery over their health needs, are experienced in dealing with professionals and formal agencies, and are ready to negotiate assertively to obtain good care." Try telling that to the residents of central Harlem, where the infant mortality rate a few years ago was more than twice the citywide average, and where the overall mortality rate is 50 percent higher than that which prevails for the rest of New York City.

Eli Ginzberg has also commented on the phenomenon of provider dominance, calling on health providers, both hospitals and professionals, to be "more alert to the needs and desires of those whom they treat," and to recognize that "the quality of health services not only involves professional judgment but also requires consumer approval."

To fully appreciate Ginzberg's theory, one must have heard a recent skit by Bob and Ray on National Public Radio, in which a hospital administrator is heard sternly upbraiding an orderly for helping a little old lady use an elevator marked "For Staff Use Only" to visit her critically ill son on the 18th floor of the hospital. It is only after the administrator's lecture is over that we learn from the orderly that all of the hospital's elevators are marked "For Staff Use Only."

Educating the Health Consumer

We need to do a better job of educating the health consumer about his and her rights in the health-care marketplace. Government has sought to do this by promulgating a written code of patients' rights in hospitals and residential health-care facilities. At a minimun, these assure that the patient will receive considerate and respectful care, know the names of those responsible for that care, and be adequately informed about the nature of his or her sickness, so he or she may assent to or refuse care.

We are proposing substantial revisions in the state's standards governing patients' rights in residential health-care facilities. Specifically, these revisions would require that patients and their families be notified of the facilities' bed retention or reservation policy; that facilities maintain patients' councils and establish a formal mechanism for handling complaints of patients and their families; that patients' accounts of accidents or incidents be incorporated in written reports; and that at least ten hours of daily visiting be permitted. The proposed regulations also set forth specific requirements for the use of medications and patient restraints, and are designed to reduce indiscriminate use of such restraints and medications that modify behavior.

To those who decry government's effort to espouse consumers' interests and regulate health-care delivery, I respond that it is private, entrepreneurial medical practice that is giving rise to the social problems that stimulate these regulatory mechanisms. Difficulties arising from the cost, access, and quality of health care come not from too much regulation, but from medical practitioners' abuse of the social contract. As Milton Roemer has reminded us, "The social physician can no longer be regarded as an ethereal ideal of scholars and dreamers; he or she has come to be expected by people everywhere."

We are approaching the end of a socially acceptable limit for health expenditures. Neither government nor the American people can afford to spend more. Thus, if we hope to continue to make significant gains in human health in the future, we are going to have to change the thrust of our health-care enterprise from one that devotes about 90 percent of its resources to curing sickness, to one that does far more than it is doing at the present to influence healthful changes in human behavior.

In order to achieve this, we are going to have to spend a greater proportion of our limited health dollars on educational efforts based on our school systems and our places of work, to convince people to adopt healthier lifestyles and practices. We are going to have to do a better job of telling the American people about the health choices available to them. This challenging task is too much for government to do alone. If we are going to change the emphasis of health from cure to prevention, we will need the input of all facets of the health care system—consumers and providers alike. We must tell the American people that the money we spend on nuclear magnetic resonance machines is money we do not have available to provide food to the hungry, to provide prenatal care and supplemental nutrition to pregnant teenagers, to replace worn-out municipal water systems, and to remove potentially deadly chemical wastes from the environment. Which do they choose?

In every ethical system there is a set of ethical values. One of the jobs of government is to arrange these values in some hierarchical order, so that when conflicts arise they may be resolved. In his collection of essays on Jewish Bioethics, J. David Bleich used the Robin Hood ethic to illustrate this value system. He writes, "Robin Hood finds himself confronted with a moral dilemma arising from two different and conflicting claims which cannot be reconciled. His obligation to preserve human life compels him to do whatever is necessary to assuage the hunger of starving widows and orphans; his obligation to respect the property rights of others restrains him from expropriating for this purpose any object of material value under the jurisdiction of the Sheriff of Nottingham. What is required is a ranking of values so that the moral agent may be guided in his conduct and enabled to preserve or promote the higher moral value. Robin Hood's conduct is predicated upon a determination that the sanctity of human life represents a higher moral value than preservation of property."

The moral dilemma that faced Robin Hood is the same one that we face today in the field of health. We must put the preservation of human life and the quality of life above the value we attach to the trappings of health, such as high-priced medical technology and multimillion-dollar hospital buildings.

Hospice as Advocacy

David Greer

Introduction

In the early 1970s, a latter-day medical missionary arrived on the shores of North America. Her name was Cicely Saunders and she came to plead the cause of the terminally ill. It was Dr. Saunders' contention that the medical profession had made dying an impersonal and technical process, and that by doing so the profession was poorly serving the terminally ill and their families. Doctor Saunders contended that the dying needed a new system of care stressing palliation and psychosocial support, and she called this new system "hospice," after the medieval church-sponsored institutions for the dispossessed and disabled.

The first American hospice was established in 1974 in New Haven, Connecticut, and in less than a decade the movement had swept across the nation. As we entered the 1980s, no one knew how many hospices there were in the United States, but the best guesses were over one thousand, scattered from coast to coast and from our northern to southern borders.

In our unique American environment, hospices assumed many organizational shapes and sizes. Strong social and political forces tugged in many directions and less apparent economic forces built momentum. By 1982, the hospice movement was potent enough to pressure a parsimonious Congress into establishing hospice as a new Medicare-reimbursed system of care—an astounding feat in an era of contracting support for social and health programs.

Dr. Saunders' seed had settled in extremely fertile soil and the elements of that fertility tell us much, not only about the state of medical care in our country in the last quarter of this century, but also about the social, cultural, and economic forces at play in our nation.

The Hospice

What is hospice? The National Hospice Organization describes it as follows:

> *Hospice affirms life. Hospice exists to provide support and care for persons in the last phases of incurable disease so that they might live as fully and comfortably as possible. Hospice recognizes dying as a normal process whether or not resulting from disease. Hospice neither hastens nor postpones death. Hospice exists in the hope and belief that, through appropriate care and the promotion of a caring community sensitive to their needs, patients and families may be free to attain a degree of mental and spiritual preparation for death that is satisfactory to them.*

This translates into the following services:

1. Health Care and service are provided to terminally ill patients and their families.

2. The patient, family, and other persons essential to the patient's care constitute the unit of care. The "hospice patient's family" refers to the patient's immediate relatives; persons with significant personal ties may be designated as the patient's "family" by mutual agreement among the patient, that person, and the hospice organization.

3. Inpatient and home care services are closely integrated to ensure continuity and coordination of care.

4. Care is available seven days a week, 24 hours a day.

5. Care is planned and provided by a medically supervised interdisciplinary team composed of several persons with appropriate skills. The team members work together to plan and provide services that will secure the physical, emotional, and spiritual welfare of the patient and his or her family.

6. Palliative and supportive care is directed at allaying the physical and emotional discomfort associated with terminal illness.

7. Bereavement services are provided; they may include follow-up visits and support of the family after the patient dies.

8. An educational program that has two components is available. These consist of: (a) Educating the patient, family, and interdisciplinary team concerning death and dying, and (b) teaching the family to care for the patient in the home.

9. Volunteers play an important role in the provision of care.

Expanding Role of Hospice

What is responsible for the astounding expansion of this movement in the United States at this time in our history? To answer this question, we must digress briefly for an examination of the American milieu into which hospice was inserted.

In the past several decades, there have been changes in the demography of the dying and the duration of the process. More of the dying are older, and more frequently, dying results from extended chronic illness. As a consequence, familial constellations and relationships to the dying patient have changed.

Dying has shifted from the home to hospitals and long-term care facilities. Coincident with this change in location, the dying process has been dominated by professionals, and technology has been increasingly apparent in the environment. Simultaneously, there has been widespread disillusionment with technology in the society. Industrial development, nuclear energy, robotics, and so on, which once inspired visions of a Utopian society, have in the opinion of many brought instead pollution, industrial squalor, dehumanization, and the threat of nuclear annihilation. The medical profession has not been immune to this antitechnologic sentiment, as is evidenced by the considerable public criticism of "overutilization" of technology by physicians.

Whereas many persons once accepted suffering in the expectation of after-life rewards, recent trends tend to focus on the good life, here and now. Materialism, sometimes extending to hedonism, is a dominant force in our current society. Medicine has always been looked to for relief from pain and suffering, but recently, improved "quality of life," as well as good health, have become part of the medical mandate.

Beginning with the publication of Kubler-Ross's *On Death and Dying* in 1969, the dying proces has emerged from the shadow of public consciousness. Dying is increasingly seen as an important part of living, worthy of the attention and interest of both professionals and laymen. Taboos that once submerged the process have been progressively discarded. Dying is discussed in the media, there are professional journals devoted to the study of dying, and educational activities for both laymen and professionals are proliferating. Increased visibility reveals inadequacies and creates the demand for improvement.

Finally, Americans revere individual sovereignty, self-sufficiency, and opportunity for personal development. We are living in an egalitarian and, to a large extent, an antiauthoritarian era. Social activism and aggressive consumerism are natural byproducts of this philosophic orientation. Consumers are no longer passive recipients of products and services. They have strong ideas concerning their needs and desires, and know how to apply the pressures necessary to achieve their objectives. Increasingly well-educated and committed to self-determination, Americans are not prone to defer so quickly as in earlier times to professional opinion or expert advice.

Hospice Advocacy

The American hospice saga is probably the most successful advocacy venture in the United States in the past decade. From Dr. Saunders' initial efforts there emerged, almost instinctively, a movement perfectly designed for the advocacy role. It rapidly mobilized a broad and powerful constituency, organized on both the regional and national levels and leveraged to drive public policy, and, in the process, created a new national system of health care amply funded and widely supported. All this was accomplished in an era of constraint on health-care costs and declining support for publicly funded human-service programs. For those interested in strengthening the advocacy role, hospice may be the ultimate paradigm for success.

Advocacy has been defined as rhetoric or orchestrated action designed to alter balances of power for the benefit of a particular group. The definition excludes advocacy on an individual level, whether it be the advocacy of the lawyer for the established rights of his client or the advocacy of the social worker to obtain desired rights for the client. It is this type of group advocacy, which reached its zenith in the hospice movement, that will be the focus of the remainder of the paper.

Group Advocacy

Group advocacy has a clientele as the primary focus of its activites as opposed to the public at large, and sees as its goal the proactive identification, pursuit, and marshalling of resources for the benefit of its clientele constituency. Group advocacy may be engaged in by professionals, potential recipients (the clientele), or by disinterested persons concerned about a problem or a needy group. Each of these can bring unique strengths to an advocacy movement, and combinations of the three can be particularly potent.

The hospice movement succeeded in combining all three: the lonely, terminally ill individual and his or her family, suffering through Kubler-Ross's stages of dying; the professionals drawn principally from nursing, with less involvement of social workers and even less from the medical profession itself; and the concerned but not directly involved groups, neither dying nor serving the terminally ill—the ministers, retired professionals, industrialists, and financiers drawn from the higher socioeconomic strata in the community.

To the movement, the patients and their families brought a cry for individual rights, self-sufficiency, and self-determination consistent with social trends. Eighty-seven percent of cancer patients died in hospitals, warehoused and encumbered in a high-technology, scientific environment. Families wanted their loved ones at home or in a home-like environment accessible to them. They spoke for humane treatment, rather than medicalized treatment, within their established network of family and friends. Consistent with the broad anti-institutional sentiment sweeping the country in the 70s, they contended that we live better and die better at home.

Professional and Volunteer Advocacy

The professional element in hospice advocacy consisted primarily of dissidents. Frequently direct observers of the medicalization of the terminal period in life, they reflected disillusionment with technology and a yearning for the return of caring as opposed to curing in the terminal period. They had often seen much pain and suffering, sometimes the result of attempts to cure or to prolong life, and they were close enough to observe the futility of many of these efforts.

It was the volunteer element, however, that brought to hospice advocacy its unique strength and, ultimately, its victory. It was the inspiration of the early hospice advocates to involve volunteers at all levels of hospice activity; they made up large segments of the governing boards, they brought the skills of their occupations to the organizations, and they were directly involved in patient care. Indeed, many of the early American hospices were started and staffed by volunteers. The volunteers constituted an expression by the community of its concerns for itself as well as for the terminally ill. Volunteers spoke for the autonomy of the lay population, as opposed to the authoritarianism of the specialist, and escape from the paternal embrace of dispassionate, self-serving professionalism. Finally, volunteers became the primary force in the public sector, in the halls of statehouses and Congress, and in the powerful lobby organized for the final offensive.

Effective advocacy ultimately requires organization, and hospice advocates understood this early in the development of the movement. In less than five years, a national organization was developed. Shortly thereafter, an "educational" arm of this organization was developed to coordinate the lobbying function. Skilled public relations and legal personnel were recruited to staff and serve as consultants to these groups. Although there was minimal information available on the cost or effectiveness of hospice care, statisticians and evaluators were employed to cull the field for the most favorable data.

The final offensive occurred in the 96th Congress in the summer of 1982, and victory was achieved in the Tax Equity and Fiscal Responsibility Act that made hospice Medicare reimbursable. Skirmishes still continue on the regional level with Blue Cross, Medicaid, and other third-party payers, but with the major battle won, ultimate victory for the hospice forces is assured. Hospice is now so well established that we are witnessing the entry of the for-profit sector into the field. The final stamp of permanency—certification, licensure, and regulation—is well underway.

When Dame Cicely Saunders arrived on our shores just eleven years ago, she could not have imagined what she and American advocacy would spawn; not only a large, national human service system but in addition, in all likelihood, a new American industry. Advocates, both academics and practitioners, have much to learn from this remarkable movement.

Health Advocacy Among Minority Groups

George I. Lythcott

Introduction

The problem of health advocacy among minority groups centers around the providers of health care (a) drawing attention to the magnitude of the problems of disease affecting the different minority groups in the United States, and (b) more importantly, planning, developing, and implementing programs that will diffuse specific information throughout these groups to assure optimal access to the health-care system and the appropriate use of that system to reduce the individual and collective morbidity, as well as the number of preventable deaths.

The hoped-for result is not simply the dissemination, or even the appreciation, of new knowledge or health information among these groups, but rather to create additudinal changes so that new information becomes credible and acceptable, even if it collides head-on with specific traditional beliefs and lifestyles. This is no easy undertaking; it is complex and confounding for a variety of reasons—but I believe, it is attainable!

Let me first establish the numbers and kinds of minorities about which we speak, then I will present significant comparative data to identify the current abysmal health status of these minorities—both reasons

27

enough for developing and improving the diffusion of information that will support the minorities access and appropriate use of the health-care system.

Ethnic Comparisons of Health

According to the 1980 census, about 46 million Americans are nonwhite: 14.6 million Hispanics, 1.4 million American Indians. 3.5 million people of Asian/Pacific island ancestry, and 26.5 million Blacks.

Now let us look, by race, at four selected parameters often used in determining health status in a community or nation—infant mortality, life expectancy, maternal mortality, and the percent of live births with no prenatal care or only third-trimester prenatal care.

Figure 1 reflects available data on infant mortality by race. Hispanics are not included in the racial breakdown and this kind of anachromism is unfortunately a

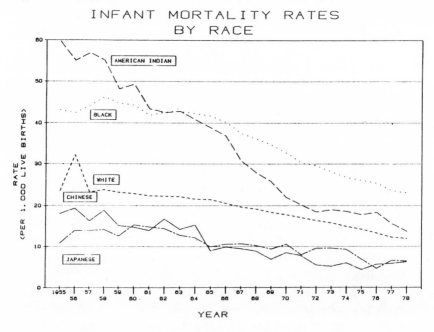

Fig. 1. Infant mortality rates by race.

common finding when attempting to correlate health-related data by race. In 1978, and this could be extrapolated to 1982, infant mortality among blacks was about twice that of whites, whereas Chinese and Japanese infant mortality in the US is actually lower than among whites. There is a precipituous decline in infant mortality among American Indian between 1955 and the present, such that white and American Indian infant mortality rates are now at about the same level. There is a lesson, I believe, to be learned from the rapidly declining infant mortality rate among American Indians, which I will share with you later in this discussion.

Figure 2 reflects life expectancy for whites and nonwhites. The average length of life in years has been, and continues to be, greater for whites than for nonwhites, although the difference in the average lifespan between the two groups has been narrowing since 1900.

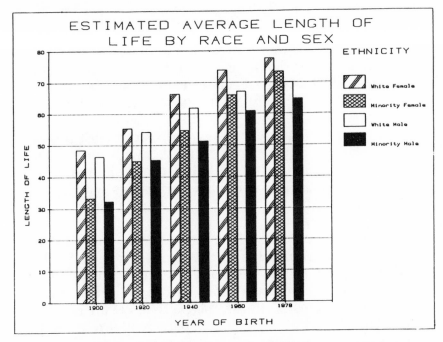

Fig. 2. Estimated average length of life by race and sex.

George I. Lythcott

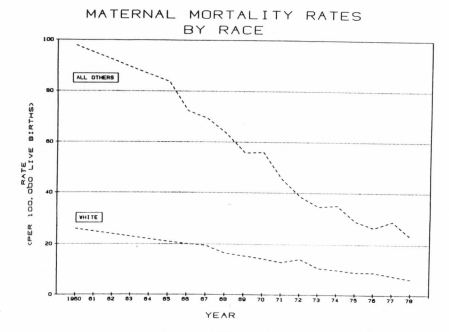

Fig. 3. Maternal mortality (whites and all others).

Figure 3 shows a comparison of maternal mortality rates between whites and "all others." Even though maternal mortality has been decreasing for both groups since 1960, "all other women" have a maternal mortality rate of almost three times that of white women.

Figure 4 compares whites and blacks with respect to the percent of live births in the US with no prenatal care or care only during the third trimester. Again, the percent of black women delivering babies with inadequate prenatal care is in considerable excess, being about twice that of white women.

The data on these four figures were compiled by the National Center for Health Statistics (NCHS), the official agency of the Federal Government for all such vital statistics. You may recall that Fig. 1 on mortality compares whites, Chinese, blacks, American Indians, and Japanese, but excludes Hispanics. Figure 2 on life expectancy compares whites and minorities; Fig. 3 shows a comparison of whites and "all others" with re-

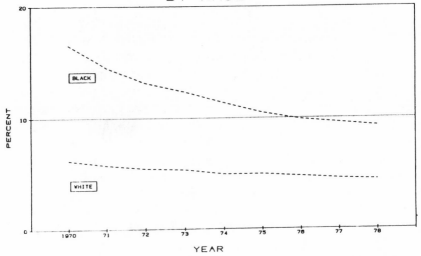

PERCENT OF LIVE BIRTHS
WITH NO PRENATAL CARE OR ONLY
THIRD TRIMESTER PRENATAL CARE
BY RACE

Fig. 4. Percent of live births in US with no prenatal care (by race).

spect to maternal mortality; and Fig. 4 compares whites and blacks. The point I am makng is that our data base for ethnic comparisons in health is faulty. There is no rational common denominator for separate minority groups, which is indicative of the lack of importance of these data to statisticians, even at the federal level. For example, let me quote from a report (Development of Diffusion Strategies Among Culturally Diverse Populations, Arrow Services Division, Mandex, Inc.) that I will discuss in more detail later: "Among people of Asian/Pacific Island ancestry, there are no national statistics showing the breakout of disease prevalence by country of origin." And from another place in the report, "The white–nonwhite distinction is further complicated by the fact that Hispanics are sometimes included within both groups according to the skin color and self-reported decisions about their racial origin."

The second point I would make relates to the cultural diversity among minorities of different ethnic groups. Although all racial minorities have traditionally been, and are still, placed in a single category—"nonwhite," "all others," "all minorities," or some equally all-inclusive definition for health-data reporting purposes—ethnic minority populations in the United States are not only highly diverse but are often proud of their cultural differences.

Furthermore, each identified minority population consists of numerous culturally diverse groups with differences resulting from different languages, national origins, cultural heritages, socioeconomic characteristics, and living habits and occupations, e.g., reservation Indians vs urban Indians. It is impossible, I submit, for an effective program for diffusing health information and education to address all of these groups, using the same techniques and methods. And the greatest of all errors is to assume that if it is effective and successful for the diffusion of information among white Americans, that it will be effective and successful for "all others."

Take, for example, cardiovascular disease, a group of conditions that ranks first in causes of death, limitation of activity, and worker disabled benefits, second in physician visits and days of bed disability, and third in lost workdays. Coronary heart disease (the predominant form of cardiovascular disease) is the number one killer in the United States. Although the prevalence of cardiovascular disease is still high, there has been a slowly declining rate since 1950, a more sharply declining rate since 1963, and a precipitious decline since 1970. This decline in mortality has been attributed to research and educational programs leading people to alter their behavior in ways that decrease risk ractors for cardiovascular disease.

The decline in mortality observed for the general population has not been reflected in most ethnic minority populations over this same period. The prevalence of high blood pressure, coronary artery disease, and stroke, as well as high levels of blood cholesterol, remains considerably higher among most ethnic minority groups than in the general population. Although there is probably no single reason for this discrepancy, it is likely that some of the reported differences in health status for minority populations are a result of the limited effectiveness of past information diffusion efforts, developed with insufficient consideration and understanding of the cultural diversity among America's minority populations.

If we as health care professionals have more than a rhetorical interest in and are serious about our efforts to improve the health status of all minorities, then we must have comparative data on all these groups by ethnic minority to be able to tell from where we have come, and to identify trends, measure progress, or target the "soft" spots that need additional emphasis or attention. Further, we must be prepared to walk that extra mile to familiarize ourselves in some detail with the cultural backgrounds of the ethnic minority groups in the communities that we serve. Evidence abounds and continues to grow, since Everett Rogers' monumental studies beginning in 1971 on the diffusion of information, that the successful dissemination of health information to ethnic and cultural minorities requires a precise knowledge of those individuals' cultures and requires the development and implementation of strategies derived from, planned around, and executed on the basis of such precise knowledge.

Cultural diversity (the differences in the social system in which the diffusion of information operates) can either facilitate or frustrate the process of diffusion. These effects are often not anticipated by an outside change agency, which is largely the result of the change agency's ignorance of certain cultural characteristics of the target population.

Presented here is one of the more elucidative examples, from the Mandex report quoted previously, demonstrating that cultural differences may lead to dif-

ferences in perception of the severity of a particular disease process, thereby reducing the relative advantage of
a recommended treatment.

> *The importance of understanding how a people per-*
> *ceive [sic] the relative consequences of illness is illustrated*
> *by the Navajo Indians of North America. Among the Na-*
> *vajo Indians there is a high incidence of congenital hip dis-*
> *location, 1090 per 100,000. The condition can be treated*
> *successfully by nonsurgical means during the first couple*
> *of years of life, and surgically during the next several*
> *years. But at later ages, "freezing" the hip, which elimi-*
> *nates motion, is the only way to prevent probable painful*
> *arthritis beginning at the age of 40–45. This is the ac-*
> *cepted treatment in American culture, and when medical*
> *workers discovered the high Navajo incidence, they as-*
> *sumed it was the answer to the problem among the Indi-*
> *ans. However, among the Navajo, the condition was not*
> *considered to be really serious. It was actually considered*
> *a blessing, since fate, having dealt this minor deformity,*
> *was thought less likely to strike a family with greater ill*
> *furtune. The congenital hip is not a barrier to marrying*
> *and having children. But, for a Navajo, a frozen hip is a*
> *serious problem. He cannot sit comfortably on the ground*
> *or on a sheepskin to eat with his family, and he cannot*
> *ride horseback. These handicaps more than offset the*
> *thought of disability 20 or 30 more years in the future;*
> *disabilities that may not come anyway. "From the view-*
> *point of the Navajo, the sole contribution of modern medi-*
> *cine to the question of congenital hip disease was to trans-*
> *form something that was no real handicap (and was*
> *almost a blessing), into something that represented a very*
> *serious handicap, indeed."*

Health Information Diffusion

We do not have the time or space to dissect the details
of the several theories of the diffusion of information.
Research and information abounds on this critical
subject—some relevant and some not so relevant to the
discussion at hand. The most relevant and, in my view,

the definitive study in our time, aimed at the methodology and techniques used in diffusing health information to culturally diverse populations, is a report published in 1983 by the Anrow Sciences Division of The Mandex Corporation, a study commissioned by the National Heart, Lung and Blood Institute of the National Institutes of Health. It is a report upon which I have relied extensively for this paper, and one that I heartily recommend.

The National Heart, Lung and Blood Institute has been involved in various efforts, programs, research, and demonstration projects over the past 15 years, using panels of experts, especially from several minority populations. The program staff of the Institute has developed, or caused to be developed, the definitive information on techniques for the diffusion of patient information into minority populations. They have done this in pursuit of their successful programs for the prevention of cardiovascular disease, hypertention, and stroke. This was done originally through several separate programs that were later combined and coordinated and that have run concurrently over the past 12 years: (1) The National High Blood Pressure Education Program in 1972, (2) The 1975 Minority Forum on Hypertention and the Initiation of a National Minority Task Force on Hypertention, (3) The Black Providers Task Force on Hypertention created in 1977, charged with providing expert advice to The Institute and its national programs, and (4) the publication in 1983 of the Report on Diffusion Strategies among culturally diverse populations, mentioned earlier.

The aforementioned coordinated effort was aimed at the careful development of a plan for the investigation and analysis of information in a variety of areas. The effort, among other things, assessed current programs and health problems, examined minority provider efforts, along with those of the federal government and voluntary agencies and local providers, studied feasible interactions and relevant barriers, and developed proposals and recommendations for improvements—the details of which are contained in the Report on Diffusion Strategies among culturally diverse populations.

The recommendations and techniques contained in the report were intended to be used not only the Heart, Lung Blood Institute, which commissioned the Report, and other parts of the federal government, but also by health-care providers themselves, representatives of large-scale delivery systems, financing organizations, and various community organizations. The recommendations were intended to be liberally broadcast in the health-care industry. But, like most other innovations, especially those that interrupt the "status quo," I am not at all sure that review and examination of these strategies in the health care industry have been carried forward sufficiently to reach the level of visibility they deserve.

Model of Information Diffusion

I submit, then, that a definitive model exists, and although created to target a specific disease entity—cardiovascular disease—the research, technology, and strategies can, with appropriate and indicated effort, be modified as necessary to support the dissemination and diffusion of health information on other specific disease states that may affect particular minority populations. For example, one of the basic but subtle truths that has emerged from these research and demonstration programs by the National Heart, Lung and Blood Institute is the conclusion that diffusion strategies aimed at ethnic and cultural minority groups are likely to be more effective if they are planned and implemented at the local or core community level.

Consultant panelists from each of the four target minority populations indicated that there is suspicion and a questioning of the motives of outside change-agents on the part of many individuals in minority care communities. This is caused, at least in part, by the residual effects of previously experienced discrimination. Consultant panelists indicate that it is also a result of the failure of past programs that were planned and implemented without sufficient community involvement.

Such concerns on the part of minority communities are by no means new. I recall this as a fundamental issue when I worked at Columbia University's College of Physicians and Surgeons in the late 60s and early 70s to improve the health status of residents in the Washington Heights and Harlem communities, which we served through our affiliating hospitals. The difference now is that providers have learned to appreciate and even respect the contributions and opinions of core communities in the satisfaction of community heads. With the growing political awareness in minority communities over the last couple of decades has come an increased pride in the community and a desire for self-determination. The American Indian communities were even able to convince the federal government of this important goal among its people several years ago. It is clear that minority communities do not want to be told by an outside change-agent what their problems are and how to solve them. What they do want is funding and technical assistance to identify their own problems and develop their own culturally relevant methods for solving them.

This is not to say that the core communities' needs are best served by "giving them the money and letting them run," for the second basic truth evolving from the new diffusion strategies is the importance of a collaborative effort—a partnership, if you please, between the core community and the larger communities. This larger community includes federal, state, and local government agencies, as well as nongovernmental voluntary health, and other organizations among whom there is the responsibility for conducting health and biomedical research and planning, and administering

primary health care, nutrition, and education programs. Clearly the core and the larger communities are
interdependent.

And, although it is true that the larger community
usually controls the bulk of available financial and human resources, and possesses greater knowledge of the
whole picture and a specific ability to make things happen, what it often lacks is commitment. And in my
judgment, it is only from commitment that all else
flows:

Summary

In summary I will restate the key issues and strategies
related to improving access and the appropriate use of
the health-care system for 46 million Americans who
happen to represent several distinct and diverse minority cultural groups:

1. We need a rational and accurate data base on all
   cultural minorities in the United States, and because of the diversity among them, we must
   cease and resist the use of all-inclusive terms
   (minority, all other, nonwhite) in the collection
   and interpretation of vital statistics, in order to
   recognize and react through appropriate strategies to the important cultural and other diversi-
   ties among them.
2. For those in the larger community who continue
   to be oblivious and who need a valid "cause,"
   we must pinpoint the considerably poorer health
   status of the several minority groups, using accepted national and/or international criteria for
   comparison.
3. Basic information, strategies, and recommendations already exist to support the diffusion of
   health information into all US minority communities. Although these were developed to support a particular disease state (cardiovascular

disease), they need only be carefully reviewed and modified to support, as appropriate, other specific diseases that affect one or another minority group.

4. Diffusion strategies aimed at ethnic and culturally diverse minority groups are considerably more likely to be successful if they are planned and implemented at the community level, not at city, state, or federal levels.

5. The respective roles of the core community and the larger community must be appreciated and supported, to the extent that a partnership between the two is developed.

6. Commitment on the part of the larger community is critical to success and, in my view, is the determining factor in the ultimate success of most strategies aimed at improving the core community.

Advocacy for Children

Sara Rosenbaum

Introduction

Case History 1

When she was eight months pregnant Marsha Smith of Abingdon, Virginia was rushed to Johnston Memorial Hospital in pain. She had not been seen by an obstetrician. There are only two obstetricians in the county and both demanded $650 in advance before they would see her or deliver her baby. She and her husband could not afford medical care.

The first time she arrived at Johnston Memorial she was told she could not be admitted without a personal physician. After a legal services attorney intervened, the hospital admitted and treated her for an acute kidney infection. She was released the following day.

The next day, Mrs. Smith returned to the hospital again in acute pain, sure that she was in premature labor. The Chief of Emergency Services refused her admission saying, "She knew she was not supposed to return to that hospital." The physician then chased Mrs. Smith and her husband out to the parking lot, threatening to call the police if they did not leave. The doctor told her that he would not admit her, even if she was in labor.

Mrs. Smith drove twenty miles across the state line to Bristol Memorial Hospital in Bristol, Tennessee. She was hospitalized for five days with a major kidney infection. Remarkably, she delivered a healthy baby a month later at Bristol Memorial Hospital, which bent the rule against admitting out-of-state residents.

Case History 2

Jean Walker, in labor, was rushed one night in early January to a small south-central Tennessee hospital along the Tennessee–Alabama border. Her husband, Edward, was a day laborer who earned just enough money to disqualify them for public assistance. The family had no health insurance.

Mrs. Walker had no physician because both obstetricians in their county wanted $400 for delivery. Mrs. Walker arrived at the hospital in the middle of the night amidst an ice and rain storm. The hospital staff admitted her to the labor room and supplied her hospital clothing. A few minutes later, the nurse came back and told her she was sorry but she would have to leave because she had no doctor. Mrs. Walker dressed and returned to the waiting room, still in labor. The nurses reconsidered, readmitted her to the labor room, and undressed her again. The nurses contacted the two local obstetricians in town, who both refused to deliver her because it was late at night, the weather was bad, and she had no money. The nurses told Mrs. Walker to get dressed again. They told her they were sorry but that she would have to go elsewhere. The Walkers drove 35 miles through the storm to a hospital in Huntsville, Alabama where their baby was delivered.

Case History 3

Frank and Ella Hogan brought their baby to the Ross County Medical Center in Columbus, Ohio, for care. Their desperately ill baby had been examined by a physician at a public clinic in Pike County who immediately referred to Hogans to Ross County so their baby could be admitted and treated. The Hogans were indigent and had no doctor of their own.

Upon arriving at Ross County, the Hogans were kept waiting in the emergency room for four hours. The baby was finally admitted by a radiologist after the pediatrician on call had refused to admit or treat the baby. The baby died a few hours later without having received medical attention other than that provided by the radiologist.

When asked why he refused to admit the baby, the pediatrician said that he was not going to serve a backup to any "free clinic." This physician appeared to have a history of refusing to admit indigent patients.

These stories are all true and represent just a sampling of those reported to the federal government each year under the Hill Burton program. Thousands more families like the Hogans, the Walkers, and the Smiths suffer similar denials of access to basic health services that are never reported. Indeed, it is difficult to classify American's health policies toward children as anything less than a national tragedy. Of all populations at risk in America, poor children suffer from perhaps the greatest inequities that the American health-care system has to offer. These inequities are all the more tragic because children are the future of this nation and because so many of the risks and illneses than plague poor children can be prevented through sound health and family support policies and a modest investment of funds.

The Children's Defense Fund (CDF) provides long-range and systematic advocacy on behalf of the nation's children. Through research, investigation, and analysis, we document many of the major problems affecting poor and minority children and attempt to remedy these problems through a range of strategies, including legislative and administrative advocacy, litigation, public education, and research. It is an especially challenging task to represent children. They do not vote. They often cannot express their needs or desires, and their best interest must frequently be measured and evaluated by others. Those who engage in the representation of children must therefore not only become experts in the problems that children face, but must make a special effort to be accountable to a population with special needs that cannot speak for itself. The purpose of this presentation is to introduce the key health problems facing children and to define the elements of effective child-care health advocacy.

Why Be Concerned About Poor Children?

Over the past half century or so, America has made remarkable progress in improving the health status and life expectancy of all its citizens. Infant mortality rates today are 40% lower than they were 15 years ago. Certain communicable diseases have been virtually eliminated, and many health problems that would have meant a lifetime of handicap and early death are now treatable.

Yet, we have major cause for concern about poor children. Over 13 million poor children live in America today—a 31% increase since 1979. This increase in the poverty rate represents the sharpest poverty rate increase for children since poverty statistics have been collected. By almost any measure, moreover, poor children are in worse health than their wealthier counterparts.

Poor children are nearly twice as likely to be born at low birth-weight (a condition that increases by 20 times the likelihood of death during infancy), twice as likely to contract illnesses such as bacterial meningitis, three to four times more likely to be inappropriately immunized during the preschool period, two to three times as likely to contract illnesses such as rheumatic fever, two to three times as likely to suffer hearing problems, 50%

more likely to have uncorrected vision difficulties, nine times as likely to have elevated blood levels, and 75% more likely to be admitted to a hospital within a year.

Poor children have 30% more days of restricted activity and lose 40% more school days because of illness. Their parents are more likely to report them as suffering from a chronic condition. Three to six times as many poor children are likely to be reported in fair to poor health, and poor children are 40–50% more likely than nonpoor children to be found to have a significant abnormality upon physical examination by a physician.

Mortality among children is significantly related to poverty. Neonatal mortality is 150% higher among poor children. Postneonatal death rates are 200% greater. After the first year of life, poor children are one and one-half to three times more likely to die then nonpoor children. Perinatal problems, when they do occur, have a greater impact and more sequalae in poor children, and poor children have great IQ deficits when born at low birthweight than other children.

Heightened Health Risks

There are indicators that over the past several years health risks facing poor children have heightened:

- Babies born to mothers receiving late or no prenatal care (in the third trimester or none at all) are three to four times as likely to die in the first year of life. Yet after a nearly 10-year period in which an increasing number of women began prenatal care early in their pregnancy, this trend has reversed since 1980, and there has been an upward climb in the percentage of women receiving little or no care.
- In recent years there has been a decline in the percentage of preschool children who are adequately immunized against childhood disease. Less than half of black preschool children are immunized against

diptheria, pertussis, and tetanus (DPT); only 39% are immunized against polio.
- There have been reports from health officials and researchers in several cities of the growing numbers of children being seen in hospital emergency rooms with failure-to-thrive problems—including stunting, anemia, and other medical complications resulting from undernourishment. In Massachusetts, children receiving Medicaid were disproportionately represented in this high-risk population, presumably because of the extreme poverty in which they subsist.

In spite of extensively documented health risks among poor children, and the growing rate of child poverty, America has no national maternal and child-health policy that assures children and pregnant women of access to basic health services. Such a policy has been recommended for decades in this country, most recently in 1980 by the Congressionally appointed Select Panel for the Promotion of Child Health. Moreover, in its report *Securing Access to Health Care*, the President's Commission for the Study of Ethical Problems concluded that this society has an obligation to ensure equitable access to health care for all. Yet, today, we are far from that goal:

- One-third of American children in poverty, and perhaps almost 20% of impoverished women of childbearing age, are uninsured. Medicaid, the largest health-care financing program for poor children and pregnant women, reaches only 40% of those who live in poverty.
- Because there is no national health financing policy for children equivalent to Medicare for the elderly, poor children depend disproportionately on a patchwork of federal programs for the disadvantaged. Without Medicaid, for example, eight out of ten poor American children would be completely uninsured. Sixty-four percent of users of Community Health Centers are children or women of childbearing age.

Public Health Programs

Because we have never established a minimum floor of decency with respect to health care for children, poor children are at extreme risk during a period when policymakers have the task of fulfilling unmet needs and have instead sought to lessen the nation's minimal public-health commitments. Basic public-health programs have been cut deeply in the last three years. Since 1981, over 700,000 children have lost Medicaid coverage and hundreds of thousands more have been prevented from qualifying because of new and restrictive eligibility criteria. The percentage of poor children who now qualify for Medicaid is the lowest since the program was first fully implemented. For those who do continue to qualify for assistance, real dollar payments per child recipient have fallen, and the children's share of Medicaid has dropped by 20% in the past four years.

Today, in the midst of unprecedented poverty rates, the Title V Maternal and Child Health Block Grant program is funded at 26% below the amount needed to maintain services at 1981 levels. These reductions have all coincided with major structural revisions of the health-care system that are designed to remove revenues and slow the rate of growth. These reductions will make charity care—already an uncertain commodity—almost entirely nonexistent.

Effective Advocacy for Poor
Children

Key children's health programs are built on a complex maze of federal/state relationships. Moreover, those programs depend for their basic effectiveness on the way in which the nation defines, shapes, and regulates its medical and health systems. Thus, child-health advocacy, to be effective, must be conducted at all levels of government and in a variety of forums. Child health advocates must be ready and able to grapple with the most complex systemic issues in American health care. Children's health advocates must work with states and localities attempting to increase county, city, and state expenditures for poor mothers and children. They must represent children before state and federal regulatory agencies charged with translating sweeping and ambiguous federal policy into specific program requirements and directives. They must represent children before Congress on issues ranging from expansion of basic programs to laws aimed at regulating health providers and controlling costs. Any initiative that alters the relationship between users and the health-care system carries implications for children, especially those for whom the system is inaccessible.

What guides our work? At bottom, of course, is the strong belief that until we affirmatively remove the barriers of poverty, deprivation, and illness, poor children's chances for equal opportunity will be diminished. We cannot expect an economy's profits to trickle down to them. They are outside of the American marketplace.

Moreover, we believe that children's health advocates must be guided by a very broad definition of health. An adequate level of income, good nutrition, and economic opportunities for his or her family are as vital to a poor child as decent health insurance. Persons who advocate medical improvement cannot do so in a vacuum. They must be responsive to the community they purport to serve, which means expanding their efforts beyond those issues that we would classify as medical.

Furthermore, advocates must be sensitive to a poor community's needs from a health-care system. What good is an excellent maternity program if the clinic is not open evenings and weekends, if transportation is unaffordable, or if there is no child care for an expectant mother's other children? What does a first-rate neonative intensive care unit *really* accomplish if an enriched, community-based follow-up and outreach program is not also provided for high-risk infants? How truly useful and effective can we expect the most comprehensive and targeted state maternal and child health plan to be if no one has consulted the families whose needs are allegedly targeted in the plan?

Summary

Despite its difficulties, child health advocacy can be and has been extremely effective. In the past number of years, CDF and many other organizations, through thorough research, public education, and network-building, have made the American public aware of the importance of considering the needs of children in the development of health policy. Although we are in the midst of an extremely difficult period of deep retrenchment, we have held out against the worst cuts. We have persuaded Congress to restore some of what was cut from key children's health programs. We are working with states across the nation on programs to strengthen and expand maternal and child health programs. We should not and cannot despair: We simply have both too much to lose and too much to gain from continued efforts on behalf of children.

Patient Advocacy

Ruth Ravich

Patient representation is a new field that has gained momentum since the late 1960s in response to increasing dissatisfaction with the delivery of health care. This momentum is evidenced by growth in membership of the National Society of Patient Representatives of the American Hospital Association. The Society, organized in 1971 with 102 members, now has some 2000 members in the national organization and its 28 local and state chapters. It is estimated that over one-half of the hospitals in the United States have a patient-representative program. The techniques that have evolved for representing patients in hospitals are now being applied in other areas, such as nursing homes, health maintenance organizations, and advocacy groups representing people with special diseases or handicaps.

This field has received continuing exposure in the media, in trade magazines such as *Hospitals, Nursing Outlook,* and *The Hospital Medical Staff,* and in books describing the patient representative's activities and tasks in improving and personalizing the patient's hospital experience. These roles and others such as acting as a referral and information resource, providing outreach to the institution's community, and serving as a catalyst for systems change, have been widely described (1,2,3,4).

In many institutions, the patient representative's knowledge of patient feelings and perceptions is used to help other health-care personnel improve their relationship with patients. Patient representative data on patient problems, concerns, and needs are used in planning for patient care and in sensitizing staff to patient perceptions. A patient representative was appointed to the President's Commission on Medical Malpractice. Patient representatives have participated in professional and lay conferences and meetings including the White House conference, "Responding to the Health Care Consumer;" annual conferences of the American Hospital Association; Mid-Atlantic Health Conferences; local and state hospital association meetings; and community meetings and health fairs. Evidence of the growth and professionalization of the field is the adoption by the Society of a "code of ethics" and the institution at Sarah Lawrence College of the first Master's program in health advocacy in the country.

There were many causes for the dissatisfaction that led to patient representation. One was a change in the early 1900s, to scientific rather than person-oriented medical education. This resulted in a higher quality of medicine, but at the same time it limited the capacity of many physicians to cope with the human concerns of patients and families. Physicians who had historically been counselors and friends, as well as doctors, were now often less able to establish a good doctor/patient relationship. The advances in medical technolgy that began during World War I and expanded rapidly during and immediately after World War II were a second major contributor to patient dissatisfaction with medical care. Physicians with access to increasing numbers of diagnostic tests and tools began to rely on the results of these to make their diagnosis, rather than touching the patient, performing a careful physical examination, and listening attentively to the patient's account of his or her symptoms and problems. This lack of personalization further breeched the doctor/patient relationship, and reduced the patient's confidence in the treatment received.

The growth in technology spurred an increase in medical specialization. As specialists, physicians began to treat specific organs, body parts, or diseases, instead of considering the "whole" patient and his or her psychosocial as well as physical needs. At present, 75% of the physicians in the United States have been trained in a specialty or subspecialty area and 60% of patients are lay or self-referred directly to them.

Specialization has also become the norm in fields such as nursing, social work, physiotherapy, respiratory therapy, and nuclear medicine. The modern teaching hospital may have as many as 250 different job classifications in 50 different occupational groups. This division of labor puts patients and their families into brief contact with, and under the care of, many different people at a time when they are least able to form new interpersonal relationships.

Television, newspapers, and magazines have played an important role disseminating information about science and health. This information, which is often well presented, helps the public learn about healthful life styles, self-care, and symptoms needing professional evaluation. But some of these presentations, particularly television programs that portray every illness as diagnosed and treated successfully within an hour, give consumers unrealistic expectations about the ability of the medical profession and often contribute to patient dissatisfaction with the available medical care.

The health-care industry, in attempting to reverse the trend toward depersonalization and the disaffection of consumers, has responded in several ways. We will discuss two of these, the issuance of a "Patient's Bill of Rights," and the growth of the field of patient representation.

The American Hospital Association issued its Bill of Rights in 1973. This document was seen as a set of guidelines to "publicly affirm the quality of care that patients could expect in health care institutions." Member institutions were asked to distribute the Bill to their patients voluntarily. Since 1973, 32 states have mandated this distribution. In New York the bill is part of the State Health Code and must be given to patients and prominently displayed in hospitals and nursing homes. The main rights addressed are:

- The right to receive the best treatment and access to programs without regard to race, color, sex, age, religion, national origin, handicap, veteran status, or source of payment.
- The right to information necessary to give informed consent prior to the start of any operation, procedure, and/or treatment. This includes the risks involved, as well as risks and benefits of any possible alternative treatment and the amount of pain and discomfort that may be expected.
- The right to know the name and function of any caretaker.
- The right to confidentiality of medical records.
- The right to personal privacy and privacy of medical programs.
- The right to refuse treatment.
- The right to know if any experimental treatment is anticipated, how and whom it will help, and that participation may be refused without prejudice to continued medical treatment.

In the 1983 revision of the Bill of Rights in the New York State Health Code, two other rights were added:

- To voice comments and recommendations for changes in policies and services.
- To voice grievances to the institution's staff and its governing body and to the New York State Department of Health, without fear of reprisal.

In some states, a patient grievance mechanism must, by law, be made available. Patient representatives have been designated to fill this role in assuring that patients' rights are upheld.

The second response to patient dissatisfaction with medical care delivery has been the development of the field of patient representation. The patient representative, also called patient advocate or ombudsman, is a generalist with a broad overview of the institution and its formal and informal rule systems. This knowledge is used to bridge the gaps in delivery of care and to enhance the specialized services that are available.

The patient representative's role is based on several models. One is the ombudsman who acts on behalf of another. This term is traditionally used in the context of providing redress for the individual in dealing with a complex governmental bureaucracy. An ombudsman is mentioned in the papers of the Ming dynasty in China as early as the 14th century, when the position was filled by an appointee of the emperor to protect the people against his advisors and tax collectors. The Swedish Governmental ombudsman, appointed by the Parliament and reporting to it, is one model used in the development of the patient representative role. He or she is characteristically impartial and has a strong sense of what is fair and just. The ombudsman has the power of subpoena, and in reporting his or her findings to the Parliament has great influence. Another advocacy model is our American system of justice, in which it is perceived that each side has a zealous advocate and that adversarial interests will advance the work of the court. Information and referral agencies, such as those developed in Britain during World War II to help locate relatives and resources for displaced citizens, is another model used in the development of the patient representative role. The crisis-oriented model has been adopted by some patient representative departments that provide patient "hot lines."

There are differences in the patient representative role and the models that we have discussed. The health advocate does not wait until other avenues of redress are exhausted, as the governmental ombudsman does, but seeks out patients in need of assistance as problems surface, thus reducing distress and formulating solutions before issues become unresolvable. In contrast to the legal advocate, the patient representative seeks to resolve problems in a nonadversarial manner, to the patient's best medical and psychosocial interest, as well as to his or her wishes. The representative's role in this area is to assure, to the best of his/her ability, that sufficient information concerning the patient's medical condition, suggested treatment, or available alternative treatment is made available so that an informed decisions about care can be made. Most patient representatives deal with difficult and complex interpersonal patient/staff and systems problems as well as the "hotel-type" services, which are also very important in health care institutions.

In addition to the protection of patients' rights discussed above, the patient representative advocates on the patient's behalf by personalizing services, negotiating the complex bureaucracy, and assuring that available services are rendered, as much as possible, to the patient's satisfaction. The patient representative establishes links between services to keep patients from falling between the cracks of interdepartmental procedures. He or she opens communication between patients, their families, and health-care staff, and serves as patient educator and acts as an information and referral source for patients, staff, and the hospital's community. Patient representatives sensitize staff to patients' perceptions of services and attitudes. They use patient complaints, problems, and concerns to recommend changes in hospital policy and procedures that will make the institution more responsive to the needs of its patients.

The patient representative seeks out those individuals in need of help, interviews patients and staff, analyzes the problems, suggests appropriate solutions, and when these are accepted, sees that they are implemented. Although assuring that patients are informed and understand regulations and procedures and that staff is aware of special needs and perceptions, the patient representative is also an important contributor to institutional planning and change.

One question that surfaces in any discussion of patient representation is whether an employee paid by the institution can truly represent people who are having difficulty within the system. I think that *only* from this position can appropriate assistance be given in a timely fashion, before problem situations escalate. Because the patient representative within the institution has ready access to the patients, their families, staff members, and administration, as well as to medical records, early case finding by the representative allows problems and concerns to be investigated and solutions proposed and carried out before a dissatisfied and angry patient is discharged. The overview of patient problems, departmental procedures, and staff attitudes that are developed by the patient representative working within the institution enables him or her to act as catalyst and change agent.

During the last few years, several important responsibilities have been added to patient representative job descriptions. In risk management/quality assurance teams, the patient representative investigates possible liability and presses for fair compensation for the patient; acting as a sensitive thermostat and pinpointing areas that need investigation, the representative makes the hospital more efficient and safe. Patient Representatives work with Institutional Review Boards that are mandated by the Federal government to review all research protocols involving human beings in institutions receiving federal funds. They must assure that the research is as safe as possible and that patients will be able to understand the purpose and the procedures before signing consent forms. These boards have both professional and lay members. Because of the scientific knowledge of the professionals, the lay members are often intimidated and may find it difficult to formulate questions that must be addressed. The patient representative, sensitive to the various educational levels of the lay members and of the prospective patients, can translate the technical information into lay terms and encourage participation of the community members.

Patient representatives are becoming active in the organization and function of Medical Ethics Committees, which deal with issues such as nonresuscitation orders, removal of life-support equipment, brain death, informed consent, human investigation, and living wills.

Patient satisfaction questionnaires are another tool that helps management stay close to the consumer. These are often distributed, collected, and collated by the patient representative department. Direct responses can be made appropriately and quickly. Information about perceptions of service and specific comments can be used to point out areas that are not serving patient needs efficiently and effectively.

In some facilities the patient representative is responsible for the provision of equal access and services for the handicapped. For instance, special care for visually and hearing-impaired patients has been or is being implemented at many institutions. Beth Israel Hospital in New York employs a full-time, sign-language interpreter. At The Mount Sinai Medical Center, an interdisciplinary team is working on methods to identify and communicate effectively with patients with hearing disabilities.

The new prospective payment plan, based on Diagnostic Related Groups (DRGs), which is being instituted by the Federal government for all Medicare patients, pays a specified amount for a particular illness, regardless of the number of days of hospitalization or the tests performed. Other third party payers are also looking into similar payment plans. With these plans it will be to the institution's benefit to discharge each patient as quickly as possible. Patient representatives are in a position to point out bottlenecks to service that prolong patient stays, to arrange for more efficient scheduling of procedures and additional consultations for patients who feel they are being discharged before they are ready to care for themselves. They can also act as liaison for the patients who move from in-patient to out-patient care and followup.

Health care has changed dramatically over the past 25 years from a voluntary effort to a big business. The operation of tremendously expensive and complex health care facilities can no longer depend on contributions from wealthy trustees and the voluntary efforts of physicians. With the increase in competition the health-care product—patient services—must be marketed, but because health care deals with sick people and their anxious families, marketing must be more than merely slick advertising and catchy slogans. Marketing must be guided by a realistic assessment of the patients' and the communities' needs and perceptions of quality and service. The trained patient representative is in an ideal position to provide planners with such assessment and to help with the development of programs and services responsive to consumer needs and effective in combining high-tech with high-touch.

We have discussed some of the issues that are in the forefront in the mid '80s. The acceptance of the patient representative role by other health care professionals is no longer an issue, and much progress has been made in broadening the patient advocacy field. During the next decade, with the increase of financial restrictions and the need for competitive, innovative and patient oriented marketing concepts, patient representatives will be called upon for even greater involvement in planning for responsive, comprehensive health services.

References

1. Ravich, R. et al. (1969) Hospital Ombudsman Smooths Flow of Services and Communication. *Hospitals, J.A.H.A.* **43**, 56–61.

2. Ravich, R. and Rehr, H. (1974) Ombudsman Program Provides Feedback. *Hospitals, J.A.H.A.* **48,** 62–67.

3. Hogan, N. S. (1980) *Humanizing Health Care: Tasks of the Patient Representative,* Medical Economics, Oradell, New Jersey.

4. Rehr, H. and Ravich, R. (1981) An Ombudsman in the Hospital, in *In the Patient's Interest: Access to Care* (M. Mailick and H. Rehr, eds.), Prodist.

5. Enman, C. M. ed. (1985) *"Patient Representation in Contemporary Health Care",* National Society of Patient Representatives, American Hospital Association.

Advocacy for the Elderly

Robert N. Butler

Introduction

I would like to stress my personal commitment to the concept of human development and deep concern with the last stage of life—late life. It is extraordinarily important that we health advocates not slice the life cycle into too many unrelated bits and pieces. It is thus of great importance to bring together advocates from many groups representing difficult stages of life, and diverse groups such as those represented in this volume need a unified program to deal with the savage cutbacks in social spending that we are experiencing in this country. I am very much committed to alliances of people of all ages, of all groups, including those related to race, creed, and socioeconomic status, including even those who may be more affluent, but may nontheless be decisively disadvantaged in some respects with regard to the health-care system.

There are a number of dangers that evolve during times of austerity that I'd like to point out. We are prone to scramble and fight among ourselves. I observed the struggle that developed in Washington: on one side, those in service and advocacy, and on the other side, those committed to basic research. Sometimes those in the world of service and advocacy fail to recognize that yesterday's research made it possible to end the need for a particular service or provided the basis for a new service. For example, with the work of Enders, Weller, and Robbins, the Nobel laureates who found the means of growing polio in tissue culture, it became possible for Salk and Sabin to introduce vaccines. For those committed to provide service and seek assurance that all school children receive vaccines, we must never forget the importance of basic research.

On the other hand, there are those in research who claim that we cannot afford to "pour money down the drain" for elaborate services, and that we must put the money into research.

From my perspective, it is important that we have an appropriate allocation of each—of R & D dollars, as well as service dollars.

Budget Priorities

How did it come about that we are now struggling among ourselves, discussing health-care rationing, or triage, which is a French word originally applied to battlefield conditions. We are in a battle—a battle of the budget and social priorities. I do not think it is appropriate to consider rationing within the health-care system without also discussing the rationing of the luxuries that so many of us enjoy in this nation.

This has been the first century in which a newborn can be expected to live out a full life. It is the century of old age characterized by the triumph or survivorship. There has been no increase in the natural inherent lifespan. But in the United States there has been a 27-year gain in the average life expectancy since 1900. And this gain is almost as great as that gained from 3000 BC, the Bronze Age, until the year 1900—nearly 5000 years later. It is extraordinary what has happened in less than a century. I call it the Longevity Revolution. But, despite this remarkable triumph, there has been very little celebration. This is partly because of the economic problems, some of which have developed since the 1973 OPEC increase in oil prices. But undoubtedly other reasons operate as well. In fact, the idealogues on the right found it very attractive to mount their particular case in the presence of economic perturbation. Where would they be if, in fact, we were still in the era of affluence that ended in 1973? We must recognize that both governments and private enterpreneurial enterprise have failed to respond effectively to economic perturbations. But Brian Abel-Smith, the British political scientist, points out that there are many dangers in making decisions today that are based upon the contemporary economic situation and that would bind us 30 or 40 years later. He traces the historic vacillations in economics going back to past centuries and emphasizes that countries do not remain in either continuing recession or inflation.

It is time for constructive forces to be mobilized. We understand that it will take time to adapt to the increased number of older persons. We have not yet fully adjusted to the Industrial Revolution that began over two centuries ago, as seen in the slums of our cities, industrial pollution, and the problems of the separation of the family from the workplace. Yet the doomsayers, particularly in Reagan's Washington, besiege us with concerns about the Longevity Revolution: "How Can We Afford Old People" was on the cover of an issue of *Forbes* magazine.

It is very important for those of us in gerontology and geriatrics to recognize that old age is just one, albeit a very important, part of the life cycle. It would be wonderful if doctors, health advocates, and health providers would join together in advocacy protective of all citizens, regardless of age and stage of life, in response to the cutbacks made since 1980. I am not very optimistic, however, because we doctors, in particular, have not been very far-sighted. We have been caught up in our own particular preoccupations and not able to work directly with those that we presumably serve: namely, patients.

Geriatrics, like pediatrics, cuts across specialties and goes beyond medicine to social and behavioral phenomena. Indeed, all the constituencies represented in this volume require reforms in reimbursement to move us away from the preoccupation with organs and procedures toward comprehensive diagnosis, care, and management, regardless of age or stage. But doctors of the various specialties will fight to maintain their fee schedules. However, Republican as well as Democratic members of Congress are angry at organized medicine and may "intrude" into the doctor–patient relationship by making direct payments to a "unit," like a hospital or a state, that in turn negotiates with the particular health providers.

Education and Aging

One of the great problems in the development of health care for older people in this country has been the failure of medical, nursing, social work, and other schools to introduce teaching related to human aging and concomitant social and clinical problems. Only minimal research dollars are devoted to the problems of aging. The National Institutes of Health budget is now over five billion dollars. Two percent of its budget goes to the National Institute of Aging, despite the fact that older people consume close to 30 percent of all medications and account for 30–40 percent of hospital days and 30 percent of all health costs. There has not been an appropriate allocation of research resources. Two of the most painful and distressing diseases of old age are the "silent epidemics" of Alzheimer's disease and osteoporosis. Only recently has there been even a minimal amount of research dollars provided to better understand the causes of Alzheimer's disease and even less to conduct fundamental research in bone physiology and endocrinology that will contribute to unraveling the mystery of osteoporosis.

The reason that high health costs are associated with old age is that there has been a remarkable postponement of illnesses and death, and therefore of health costs, into old age. Eighty percent of all deaths now occur after age 60. At the same time, we have had remarkable reductions in maternal, childhood, and infant mortality rates.

There are various causes of increasing health costs that are in in fact not directly related to aging. There has been an increase in inflation since 1965 and the health industry's rate of inflation has been even greater than average. Medicare has offered open reimbursement without controls. Nonprofit hospitals now come under the National Labor Relations Board; thus many hospital workers who had been underpaid for many years have begun to receive better wages. There is little agreement on what is appropriate treatment or prophylaxis in relationship to a particular condition. For example, in a hospital in Illinois the cost of prophylactic postsurgical antibiotics is markedly different from that in Charleston, South Carolina. I regret, as a physician, the physician's failure to take responsibility in exercising control over costs.

There have also been very important technological developments that have contributed to higher costs. I oppose the attitude of some that we must cut back on technology. We should separate payments for technology from payments for the cost of care. The Longevity Revolution is welcome, and working together, we will adapt our institutions to the changes it brings.

Advocacy Issues for Older Women

Myrna Lewis

This chapter will address key health advocacy issues for older women. The thesis discussed here is twofold: (1) older women's health issues are significantly different from those of older men, and (2) these health issues are moving into public prominence. In fact, they are on the verge now of becoming key political issues. I will discuss what is already in place in terms of advocacy for older women, as well as resources that currently exist, but are still relatively unexploited from an advocacy point of view.

How are older women's health issues different from those of older men? Or from younger persons of both sexes for that matter? The critical factor is the remarkable longevity of women compared to men. Old age is a new land, at least in terms of the numbers of people growing older, and older women are the majority of the pioneers in this new land. In the 19th century, the evidence suggests that women already lived an average of two to four years longer than men. In 1900, the date of our first official records, the difference was recognized as two years. By 1950, women were outliving men by six years. By 1981, it was seven and a half years, and the Census Bureau is now projecting that by the year 2050, women may live ten years longer than men. There are many things, of course, that could intervene in terms of conquest over some of the diseases of older men, but the evidence does seem to be clear that there will continue to be a longevity difference between the sexes.

A practical consequence of these figures is that in 1982 there were 16.6 million older women and only 10.7 million older men. With each year past 65, the proportion of women to men grows greater. Women over 85 are the fastest growing age group in the United States.

Why are women living so much longer than men? Some credit genetics: women are simply built stronger. Some emphasize environmental factors (e.g., men perhaps meet with more work hazards than women), and others see an interplay between these two along with additional factors, such as health practices and social roles. Many believe that smoking and excessive alcohol are major influences. This generation of older people had many men who smoked and drank more heavily than women. A majority opinion among researchers is that perhaps as much as two-thirds or more of the differences in life expectancy is related to a combination of environmental exposure, health habits, and social/ cultural behavior, and perhaps one-third is genetics. We cannot underestimate the obvious factor of the greater durability of the female body, as seen in most of the animal kingdom.

The illness and disability patterns of women follow logically from their greater longevity. Since women have a longer lifespan, they tend to accumulate both more chronic diseases and more disabilities before death. This may not be inevitable: It may merely reflect the fact that we have not yet learned enough about prevention and self-care to keep these long-living bodies in better states of health.

Let us look at some of the current characteristic differences between men and women. Older women now report both more acute and chronic illness and greater diability than older men, but—and this is important— they die at a lesser rate. Another way of saying this is that when men do report that they are ill, they are more likely than women to die. Some of this may be the result of women's greater freedom to report their illnesses, but the general consensus holds that women currently do have a great incidence and prevalence of illness. Older men, on the other hand, use more acute hospital care than women at any given age, because more of them are in the last year or two of their lives at any given point. Women use outpatient services of all kinds more than men, and older women typically have multiple-health problems requiring coordination of care among various caregivers. This, again, reflects the accumulation factor.

The cost of long-term care is one of the fastest growing items in the American health-care budget. Older women are the predominant users, primarily of nursing homes, as well as home and community-based, long-term care. Over 75% of nursing home residents are female. The reasons for this are straightforward: Women live longer and we know that the longer one lives, the higher the possibility of being in a nursing home. Second, women often outlive their spouses, so there is no one to take care of them, although they may have nursed their spouse through the last years. Third, women tend to have more organic brain disease than men, not because they are female, but because organic brain disease is associated with age. Organic brain disease is one of the main factors precipitating the move into a nursing home.

Fifty percent of women now in nursing homes are childless or have outlived their children. Twenty-five percent of all women 70 years of age, in and out of institutions, have no living children. This particular generation is the low-fertility generation who gave birth during the Depression, and many of them had no children, or only one or two. These women often outlive their children, particularly if the child was a male.

Finally, female caregivers in their 40s or 50s or older are increasingly in the workforce and unable to care for very sick relatives. And it is now entirely possible for five or even six generations to be alive in one family. With the middle generation of women shouldering most of the responsibility for such physical care, we can see that they clearly cannot always sustain this load.

The financing of health care is very different for women than for men. Older women have fewer personal financial resources for health care than older men, and these have to be spread out over a longer lifetime. This is proportionately more severe, of course, for minority women, but it is characteristic of the majority of all women. To finance health care, most women turn first to Medicare. But, how well does Medicare really fill the needs of illness and disability seen in older men and women? Medicare was designed through its reimbursement structure to encourage acute inpatient care rather than chronic outpatient care, which is more typically needed by women. For example, it excludes payment for prescription drugs, long-term home care, eyeglasses, dental services, routine exams, and foot care, and it does not cover nursing-home care.

Medicare, in fact, only covers 40% of the average health-care bills of those over age 65. The rest must come from personal funds or private health insurance. The amount of funds for older women, in terms of total median income, was just a little over $100 a week in 1982, or slightly over $5000 a year, compared to somewhat over $9000 a year median income for men. Forty-four percent of black older women are living in poverty. Twenty-seven-and-a-half percent of older women of Spanish origin and 16% of white older women live in poverty. Although minority women are the poorest of the poor, the vast majority of the elderly poor are older white women. The median annual income mentioned above is just $700 over the poverty level, so we can see that the majority of women, if they do not already live in poverty, are not very far from it.

A unique feature of old age is that women who are poor all of their lives are joined by the newly poor—the middle class and even upper class women who can sink to the poverty level, depending upon whether they outlive their resources or whether they experience catastrophic illnesses. Few, except the very wealthy, can finance long-term care in or out of institutions. When resources run low, women typically turn to Medicaid, which is welfare, but they must spend down to the pauper level before they are eligible. If they survive their illness and recover, they then have reduced personal financial resources to live on. Furthermore, because women live longer than men, couples often have spent all of their resources for nursing-home care for the man's dying days in a nursing home. This, then, leaves the wife with few resources for her own old age. Since the average age of widowhood in the United States is 56 years, this means that 10 to 20 or more years are left to rely on Medicaid to pick up the pieces of a dependency that might have been prevented.

Much of the extreme poverty of this generation of older women flows not only out of their longevity and the inflation in health-care costs, but also from the fact that many of them are housewives whose work was not considered "real work." They seldom have private pensions or private health plans. Many could not earn social security on their own. Those who worked outside the home typically held low-paying jobs. Older women end up in old age depicted as "problems" or "drains" on the health-care system after a lifetime of work in and out of the home that often included acting as nurse for their aging and dying parents, perhaps their husband's parents, and eventually for their spouses.

This is becoming an important political issue in two ways. The entire health-care financing system for the old, namely Medicaire, is in serious trouble, as is health-care financing in general. Also, the Reagan administration favors cuts in federal spending for social programs as a way to at least appear to be doing something about the growing federal deficit. Medicare and Medicaid are and will be major targets. We shall see this come to a head as the Medicare crisis is allowed to surface as a topic for political debate.

Meanwhile, what has happened as a result of the recent cuts in Medicare and Medicaid? Sixty percent of those on Medicare are older women, as are two-thirds of all Medicaid recipients over 65. Three-fourths of the elderly receiving Supplementary Security Income and 70 percent of the elderly receiving food stamps are women. Fifty percent of all persons of all ages in federally subsidized housing are older women living alone. One would have to conclude that the military budget expansion and social service cutbacks have already disproportionately affected the income and health of older women.

The most recent poverty figures verify this effect, with older women carrying a disproportionate share of what we call the "new poverty," generated since 1979. The overall poverty rate has gone from 11.7 percent in 1979 to 15 percent in 1982. But, if you look at older women you see something quite startling. Seventeen percent of all older women are living in poverty. An even more revealing figure is that 41 percent of all older women, mostly widows, live alone, and of this group, 30 percent are living in poverty. This includes many of the very old and the most frail. There is almost no comparable category of men in this group, since their incomes are much higher and most live with their mates or in families until they die. As one older woman put it when referring to her meager Social Security check, there is always far too much month at the end of the money. Or as another person observed about the Reagan administration, "Thank God it is once again a wonderful time to be rich."

Regarding advocacy issues, older women fortunately are not alone in the upcoming battles that will be fought over Medicare and the allocation of resources.

Their children, of course, have a stake in not having the cost shifted to them. So the middle-aged have a financial as well as, we would hope, a moral interest. Many in the younger generations are at least dimly aware that they, too, will grow older and that the bed being made for our elders is the one that we are likely to end up lying in ourselves. Advocacy organizations are emerging for older women, such as the Older Women's League, and in a broader intergeneration context, the Gray Panthers. Aging organizations involving both men and women will play a significant protective role in the upcoming struggles. The American Association of Retired Persons (AARP), for example, with its 18 million well-organized members has already done an impressive job with the Social Security crisis. The National Council of Senior Citizens is another active group. Women's organizations are also confronting what has been a tradition of agism in their own organizations—the emphasis on the problems of midlife and younger women—and they are beginning to represent older women's interests as well. Social Security reform and nursing home costs are current focuses of this particular endeavor. Civil rights organizations are interested in the protection of minority elderly and the civil rights of all older people, especially women.

For an emerging rainbow coalition of groups gathering around an issue, the upcoming Medicare struggles will likely be a political topic to watch. Interest ranges from the far left with civil rights and human rights organizations, to the far right with its "save the family" concerns. The gender gap may have an impact as well. The differences in men and women's political attitudes first began appearing consistently in the 1970s. What are these differences? Women have begun to place greater emphasis than men on civil rights and the government's role in reducing the income gap. In 1980, Ronald Reagan won the male vote with a solid margin. He did much less well with women, with a gender gap of nine percentage points between men and women. That gender gap looks even more impressive when one sees that in 1980, 59 percent of women and 59 percent of men voted. Thus, this is the first time there has been an equal vote between men and women. But there are now six and a half million more women voters than male voters.

In looking at the older woman voter in the 64-to-74 age group, 75 percent are registered to vote and 66.7% voted in 1980, compared to the average of 59 percent overall. Even above the age of 75, 53 percent of these older women voted. With growing numbers and such high voter participation, I would predict that older women will soon be serenaded for their vote on behalf of their own interests. We have already seen the politicians' anxiety about the older vote in the recent Social Security crisis. In addition, three major national organizations have recently formed to register even more of the low income, the elderly, and women for voting.

There are untapped resources for advocacy purposes as well. More older women will soon be emerging as office holders in the not-too-distant future. Since 1973, the number of women in political office has tripled. This is not as significant on the national level as it has been in the state and local levels, which have seen a sixfold increase in women mayors and a doubling of top statewide offices since 1973. There has been a rise in women state legislators from 5.6 percent in 1973 to 13.2 percent in 1983. On the local level, this is even more significant. We are seeing a growing constituency of women in elective offices. And one would surmise that the interest in political office will not disappear as women get older, just as it seldom disappears with men. Women will mature into higher positions as they gain more experience. The one female Supreme Court Justice is growing older too, so we hope to hear something from her in terms of support for the interests of older women.

There are also many fortunes in the hands of older women. Although there is tremendous poverty, one of the byproducts of living longer than men is that many women have inherited their husbands' fortunes. This has never really been exploited in any way by organizations, though it holds tremendous potential. We have seen what some women can do with their fortunes. For example, Mary Lasker in New York has done a tremendous amount for basic science research. In Washington, DC, Florence Mahoney has been greatly involved in supporting the development of the National Institutes of Health, particularly the National Institute on Aging.

Time and good health are also often in the hands of many older women. Most are not employed and many have few childcare repsonsibilities, unless they are helping other generations. Most older women no longer have mates, so they are freer than any other group in the adult population. There are also new generations of women moving into old age, many of whom have been active politically and socially in the community, and who also have a work experience outside the home. Most important, however, there is a realization that the major issues for older women are really the issues of life enhancement and human rights for everybody. This means a floor of financial security that does not require humiliation and stigmatization, a comprehensive national health care system that does not deplete the resources of sick people, and emphasis on self-care and prevention of illness whenever possible throughout the life cycle.

Older women have had to endure a great deal of abuse from the health care system, including negative verbal descriptions—not only in the health-care system, but also in the health-policy system. Medical students often call them "crocks," and doctors frequently see older women as complainers and "kvetchers" and stereotype them as having postmenopausal syndromes.

They are often overmedicated rather than thoroughly evaluated, and politicians tend to depict them as burdens and drains on the health care system.

As advocates, we can help promote a new image. Older women might be enormously pleased to see themselves as triumphs of medical and public-health progress, and as models for men and the younger generations on the cultivation of traits that enhance life expectancy. Most of all, older women deserve to be seen as hardy and brave pioneers in the new land of old age, which we all one day hope to inhabit.

Unionized Health Workers

Victor Gotbaum

INTRODUCTION

About 25 years ago, I had the terrible problem of representing the only organized hospital in the City of Chicago—the *only* organized hospital. At that time, the workers were paid the magnificent salary of $1.15 an hour, averaging almost 30 cents more per hour than other hospital workers. Justifiably, the administration of the hospital felt that they were so far out in front on the salary issue that the negotiations became quite difficult. So, we chose to organize, and did so by focusing on a couple of hospitals.

To do a professional job of organizing, one must first look into the institution and define it: define the workplace, define the membership to be organized, and acknowledge that a very tough and complete job must be done. It was fascinating that we almost immediately reached the people we wanted to organize.

In reading the hospital literature, in this case from Mt. Sinai Hospital, I saw that tribute was paid to the administrators and the trustees (after all they provided a lot of the funds), and every once in a while, there occurred statements about the doctors, the new buildings, or the building drives, and in a very rare moment, something hidden away about a nurse. But, those individuals we now call paraprofessionals—the nurses' aide, the housekeeping aide, the dietary aide, the custodians—were unmentioned and remained faceless.

There was no recognition of these workers as personnel of the hospital; they just didn't exist. This was not unique to the hospital system of Chicago, but was prevalent nationwide.

Organizing Health-Care Workers

Thus, we began to organize. We signed 94% of the workers in a matter of *weeks*. We tried but did not receive recognition, and so decided to strike. The strike lasted for about four months. When you read or hear about a strike you often hear about the horrendous aspects—what it is doing to the patients and the institution—and immediately the labor leader is at fault and should be sent to jail. Terribly negative reactions are common, and no one bothers to explain what it means to the workers who are on strike. We set up soup kitchens. We slept in the back of a station wagon. We had *no* income while the strike continued for four and a half months. I had hoped, because the wage at that time for paraprofessionals was a mere 85 cents per hour, to raise it to a dollar. Those in labor relations know that a 15-cent increase (a little more than 16–17% the first year) is generous. To bar the union, the same administration that told us they lacked funds raised the wages to exactly what we were getting at the Chicago Hospital, $1.15 an hour, just to keep us out. We closed the strike down and I had to report this to the workers.

What do you say to workers who have been on the picket line for months, who want to be recognized, and are nearly in a state of depression? I was terribly apologetic and called off the strike. During a strike, you keep spirits up, almost like using a dangling carrot—just a few more days, a few more days. The strikers were magnificent. They held out. When I informed them of the results, I expected a negative response. Instead, there was a silence, and a 75-year-old woman—I still remember her walking the picket line every day—got up and said, "Victor, we want to thank you. If you hadn't come into our lives nobody would have known that we existed. So we owe you a lot." I turned my back to the group and started to cry. Nobody recognized them; nobody cared about them; nobody knew that they existed. A part of the hospital workforce was completely and totally unrecognized.

Recognition of Paraprofessionals

Later there were better and happier days. Happier in the sense that we organized and we succeeded. We are now organized, but there is still no real recognition. We talk of a team concept. We talk of working together. We want to focus on the patient, which is something we should do, but if people can't identify with the patients, they feel isolated, they are not recognized as a team, and you can't talk of a team concept. It just has no meaning. One of the first things we did in terms of unionizing—it was not *just* a question of raising wages and fringe benefits—was to prevent doctors and nurses from calling our members by their first names. If they could call our members by their first name, then our members could call them by their first names. This became a monumental victory. Lillian Roberts, who was working with me at the time and is now Labor Commissioner for the State of New York, insisted that the nurses' aides wear white, which upset the nurses terribly. There was still a demarcation and this demarcation evolved into a very big issue.

It is true we increased wages, but more importantly we obtained recognition. The one type of recognition we still have not obtained in many hospitals is that there is no identification with the hospital. There is no feeling of being a part of the team. There is no feeling that everybody is really working together toward a common goal. A hierarchical standard remains, and not just for the people we represent. Some of the angriest people are the nurses who feel (a) they are underpaid, (b) they are not properly recognized, and (c) they are

doing terribly important work. From the union's van-
tage point, and I do not want to decline responsibility,
because I think we have a responsibility along with
management, the question is how to focus on the patient, think in terms of the entire staff as a team, and
recognize that every individual has an important role to
play in administering patient care.

Involving Paraprofesionals in Patient Care

An interesting example of the difficulty of this situation
is seen with the emergency medical service. Here, a
team approach is used among ambulance drivers,
whom we represent. The motor vehicle operators be-
lieved that their primary role was to take care of their
vehicle. When the patient was in the vehicle, the driver
was totally divorced from the role of patient. What
made him or her an important person was that vehicle,
that ambulance. The operator knew how to take care of
it and that was fine. Then the hospital administration
organized an emergency medical service and introduced the retraining of motor vehicle operators to assume new skills for the care of the patient—a
magnificent idea in theory. But when speaking to the
workers I represented, those who paid my salary, they
were absolutely furious. Their piece of security was the
ambulance. No one had ever asked them to look at the
patient. They were never a part of it. And what they
wanted to know was, "What the hell are they trying to
pull?" Hours were spent explaining that this would enhance their job and add greater meaning to their role.
The interesting and enlightening aspect of this is that,
in terms of the system, they were isolated. They had
their own piece of the action—the care and maintenance of the ambulance—and they wanted no new part
in taking care of the patient, because they never felt involved. We faced a monumental task getting them to focus on what the hospital was really all about. It is now

worked out. I do not want to say that we have perfect harmony and bliss in the medical service, but the workers' titles have been changed and their income has been increased. In terms of the ambulance and the emergency service, they now understand that they are an important part of a greater service.

Overall, we still have not been able to present the concept of working together. We focused on the patient and the fact that we are all part of the system. We just have not succeeded yet. I can go into a hospital, get the feel of that hospital, and know whether or not workers care. I try not to be prejudiced. I have been in Memorial Sloan-Kettering Cancer Center and in New York University Hospital, to name but two. In one, the workers are organized; in the other, they are not. Yet, in both hospitals there is a true sense of the team approach. There is more satisfaction on the job, and obviously, attention to salary. I think that both of these hospitals have done a good job, but there is still a way to go. A terribly hierarchical structure remains, and must be dealt with.

Summary

There is so much disparity within our society and throughout our national system. This is also evident in hospitals. Some workers receive magnificent rewards, whereas others are grossly underpaid. Because I do not represent nurses, I am free to say that you can begin by organizing the nurses, and continue with the nurses' aides. There is something very sad about the entire situation, and the disparity in health care that is evident in the workforce. When you reflect upon a system that defines life and death, it is unbelievable to have to acknowledge areas of scarcity and overabundance of medical services. It makes absolutely no sense. And I think the providers also feel this. So in the union, we do not look at our members solely in terms of their individual needs. We recognize them as stark and tough parts of a changing system. We recognize that we need to be concerned about the containment of hospital costs, and we need to represent our membership in a way that benefits all members of the health-care system in the country.

Quality Management neé Quality Assurance

Vergil N. Slee

Introduction

The chief question that should concern patients, providers, and society today is this:

> *Will the price we pay for lower cost medical care be lower quality?*

This is a real danger. The only way to know, and to take preventive and corrective action, is to understand quality management and to be sure that its essential processes are in place in the health-care setting.

Definition of Quality

First, then, let me define quality:

> *Quality (of care) is the degree of conformity with accepted principles and practices (standards), the degree of fitnes for the patient's needs, and the degree of attainment of achievable outcomes (results), consonant with the appropriate allocation or use of resources.*

The latter phrase carries the concept that high quality is not equivalent to "more" or "higher technology" or higher cost. The "degree of conformity" with standards focuses on the provider's performance, whereas the

83

"degree of fitness" for the patient's need indicates that there are considerations that override strict conformity when it comes to the individual patient.

Responsibility for Quality

The corporation (hospital) is responsible for both the quality achieved and for quality management. Although acceptable quality can occur more or less by accident, the institution must see to it that quality is not hit-or-miss, but rather the result of personal performance and systematic procedures that make it difficult, if not impossible, to provide less than desired quality.

Both statutory and case law have established the fact of corporate responsibility. Furthermore, it is logical that the responsibility must be institution-wide.

20–1 Laws

There are two 20–1 laws in quality assurance, the first of which supports this position. Simply to differentiate between the two laws, I'll call them the

> 19–1 law, system–physician
> 21–1 law, people–things

19–1 Law (System–Physician)

> *The 19–1 law states that when the care provided is not what the doctor really wanted, 19 times to 1 the doctor's decision was correct, but the system by which it was implemented failed in some respect.*

In the olden days, only the physician and the patient were involved in care. Today, there is often a long chain of people and procedures (including those procedures in the hands of the patient) between the physician's decision and the actual provision of care. There must be no breaks in this chain if the care provided is to be up to standard.

When care is provided in an institution, the institution's nonphysician employees usually carry out the 19 steps and the doctor, the one. Quality depends on each of these steps being carried out properly.

The Quality Function

We now speak of the "quality function" of the institution, where we used to talk of quality assessment, quality control, or some other part of the whole quality picture. The term "function" is used with regard to quality exactly as we would speak of the "fiscal function" or the "personnel function":

> *The quality function is the sum of the activities wherever performed, through which the institution achieves quality patient care and quality services in support of that care.*

Governing Body Responsibility

The body responsible for the hospital in its entirety is its governing body, the Board. The Board has four duties:

- Policy
- Resources, including personnel (managers and others)
- Oversight, to see that things are running according to plan
- Accountability

A corollary is that the Board itself does not manage, but rather provides agents to do so.

Feedback Loops

Among its policies with regard to the quality function, the Board must see to it that both quality improvement and quality control are carried out. Drawing the distinction between these two activities is an important advance in understanding quality in recent years, and worth elaboration. At the heart of quality activity is the feedback loop, which requires:

- Standards (what is desired—A)
- Information (what is going on—B)
- Response:
  If A = B, then keep it that way (maintain control)
  If A > B [if care is lower than the standards, change it (improve)]
- Continuous iteration of the loop

There are two kinds of feedback loops in the quality function.

The Quality Improvement Loop

This loop was first recognized in the 1950s and 1960s when the medical audit was introduced.

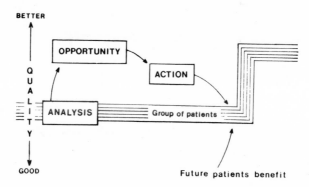

Quality improvement is the sum of the planned and managed activities through which the hospital makes specific (upward) changes in quality—creates wanted change in quality.

The use of the word opportunity in this feedback loop should be emphasized. For too long, we have had in that box the word "deficiency," a depressing word if ever there was one. It is remarkable how much more willing people are to participate in the "quality assurance" processes when they can look for opportunities to come up to their own standards, or improve on them, rather than expecting at every turn to be criticized for failures.

The Quality Control Loop

The quality control loop goes back a hundred years, and was so well-established that, although critically important, it was virtually unnoticed. In a sense, it was rediscovered in the 1970s:

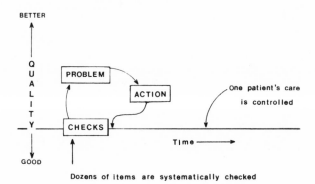

Dozens of items are systematically checked

Quality control is the sum of the planned and managed activities through which the hospital maintains quality—prevents unwanted change in quality.

Note that in both instances, the achievement of quality is the goal, not simply its measurement. (This was the important change in the quality movement in the mid 1970s—"QA" changed in meaning from "quality assessment" to "quality assurance.") The distinction between quality control and quality improvement can be emphasized by a comparison of the effects of iterations of the two kinds of loops:

QUALITY CONTROL PROVIDES

THE QUALITY <u>FLOOR</u>

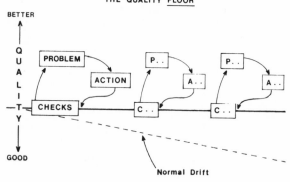

QUALITY IMPROVEMENT PROVIDES

THE QUALITY <u>STAIRCASE</u>

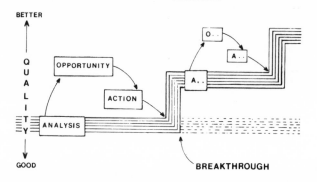

Protocols In Care

One approach to the question of quality within a fixed price is for clinicians to develop "protocols" or plans of treatment, sometimes called algorithms, for whatever unit of care the pricetag represents. Today this unit is the Diagnosis Related Group (DRG). These protocols or algorithms are step-by-step statements of the processes that should be followed under the circumstances presented by the patient.

Patient care is a process, and fortunately J. M. Juran, the dean of quality control, a number of years ago discovered the universality of the, "Pareto principle" (named after the Italian economist).

> *The Pareto principle states that in any process, there are a few vital steps, and a trivial many.*

Juran asserts that quality improvement would be impossible if this were not true.

The vital steps in a process are identified rather easily. If numerical data are available, as often is the case in the manufacturing industries, a Pareto analysis will always show that a few steps are responsible for most of the problems, whereas other problems are distributed over many steps; hence these latter steps are trivial.

This principle has been responsible for the effectiveness of quality improvement efforts in health care in California, for example, where the teaching of medical care evaluation by the state medical society limited attention to the few things: (1) on which there was professional agreement as to importance, and (2) where something could be done. Also of importance was the fact that it is impossible to address very many things at once—only a few can be handled, and, therefore, these must be the most critical.

In health care, where the data on the vital steps are often neither hard nor numerical, it turns out that professionals agree quite readily as to the most important factors. This agreement can be elicited quickly by a skillful "facilitator," for example, in developing criteria for a medical audit. Such agreement can also be found by more formal methods, such as that used by the Nominal Group Technique (NGT) developed by Delbecq (1) or other methods for quantifying "soft" data.

It is likely that we will see hospitals develop protocols for the DRGs for which they are appropriate to make sure that, in the face of limited resources, patients get the important steps of care, the omission of which would lead to both low quality and liability.

Two observations may be made:

- For a great many DRGs, it is useless to attempt to develop protocols, because they are so heterogeneous in their composition. Over 100 DRGs, for example, contain more than 100 different diagnoses. Protocols are practical only for clinically homogeneous groups of patients. (There are about 40,000 separate diagnoses codable to the 11,000 pigeonholes of ICD-9-CM. These are, in turn, grouped into 467 DRGs, so most DRGs must be mixed bags of diagnoses.)
- The development of such protocols offers one approach to eventual modification of DRGs into more clinically useful groups (or their replacement by something else). The price tag for care for a given kind of patient should be determined by the care that kind of patient typically needs. DRGs came about from the opposite direction. They were developed by trying to make somewhat reasonable clinical groupings of patients who needed roughly the same resources, as measured by dollars rather than clinical care.

Quality Management Audit

The second approach to defending quality (particularly important in the face of cost containment) is for the Board of the hospital to audit the quality management, in a fashion parallel to the way it audits its financial function. Briefly, the Board appoints a quality management audit committee, and asks it to report back in a year or so (the first time) on (1) whether the "statements" it receives about the quality are true, and (2) whether the quality is the result of a dependable sytematic approach, i.e., the result of management rather than accident.

The major differences between the quality management audit and the fiscal audit are (1) that the fiscal audit is the result of decades of experience and evolution, whereas the quality management audit is a brand new idea, and (2) that the fiscal audit has the services of a professional auditor, a certified public accountant (CPA), whereas no such person has appeared in the quality field. (The accreditation surveys of the Joint Commission on the Accreditation of Hospitals are not comparable in the quality field to the audit of the CPA in the fiscal field, i.e., they are not comparable to the quality management audit here proposed.) A QM auditor may one day appear and may even be obtained from the public accounting firm. It should be noted that such an audit will still depend on peer review for evaluation of care—it will concern itself with the quality of that peer review rather than its findings.

In order to tackle the task, the quality management audit committee must ask five questions:

1. Is there a plan for quality management that, if carried out, would work?
2. Is the hospital's organization appropriate to the task?
3. Have the necessary resources been provided?
4. Is there evidence that the plan is being carried out?
5. Is the plan getting results?

The Board, after completion of such an audit, is in a position to meet its final responsibility, that of accountability:

> *Accountability for quality is providing, to all concerned, the evidence needed to establish confidence that the quality function is being performed.*

This would be considerably easier if the Board had the credibility of the public accounting firm to point to, as it has with its fiscal audit.

21–1 Law (People–Things)

> *The 21–1 law states that the most important factor in the quality achieved is the performance of individuals, by a ratio over all other factors of 21 to 1.*

The "1" represents quality assessment methods, data sources and displays, machines, technology, and so on.

We have perfectly adequate means of measuring quality (for this purpose, estimates are just as good as scientific studies accurate to the third decimal point) and of detecting places where there are opportunities for improvement or deviations from standards. Generally, hospitals compile and have at hand the necessary information. Where we fall down is in managing so that people perform dependably, reliably, "with quality." (Actually, our data systems and "measurements" are becoming so effective that the problem often is which opportunities or deficiencies to address first—how to prioritize; here also one applies the Pareto principle.)

Only recently is emphasis being given to the difficulty of managing to maintain steady performance. Most quality is entirely adequate, and the challenge to management is to motivate people to keep things the way they are, to avoid downward drifts in quality, and to avoid innovations without proper evaluation of their likely effects—and to reward this "pedestrian" behavior. All the attention and most of the rewards are for change, yet we should reward nonchange equally when it is correct, appropriate, and efficient.

New Nomenclature: "Quality Management"

The title of this paper is "Quality Management neé Quality Assurance." There are basically two reasons for promoting this change in nomenclature:

- First, lawyers would prefer a term other than "assure," which may be interpreted as a guarantee, or may be confused with "insure."
- Second—and here is where the 21–1 law comes in— the task of achieving quality is really one of management, not of measurements, medical audit studies, or having a quality assurance coordinator.

The task of management is to get things done systematically with (through, by) people.

Every line manager in the hospital is expected to see to it that the people for whom she or he is responsible perform "with quality," do their jobs correctly, and follow procedures. This is true of the medical staff members who are expected to provide medical care in accordance with accepted standards. For the physicians, the hospital has the medical staff organization (MSO), which has as its primary duty management; that is, the carrying out of the hospital's policies and procedures with regard for the quality of care. In this usage, the chief of the medical staff is the line manager for the medical staff.

The End Product: A High-Quality Hospital

Finally, a set of eleven criteria is proposed to describe a high-quality hospital:

1. No harm
2. No errors
3. "Best" treatment, "best" care
4. Pleased patients
5. No "fat" (waste is poor quality)
6. Delighted doctors
7. Happy hospital help
8. In compliance (with regulation, accreditation)
9. Contented community
10. Not accidental—systematic
11. Beaming board

Reference

1. Delbecq, A. and Van de Ven, A. (1971) A group process model for problem identification and Program Planning. *J. Appl. Behav. Sci.* **7**, 466–491.

Confessions of a Naive Scientist

Irving J. Selikoff

INTRODUCTION

Forty years ago I was under the same misapprehension as many scientists and doctors; that our task was to simply obtain knowledge and information that could be used for human good. That makes sense. The assumption probably goes back to the beginnings of medicine, which began about 5000 years ago with a combination of magic and religion originating from the innate desire of human beings to help others. One might also offer the biological explanation for such almost innate feelings: that those who helped others survived better as a species over the years. After all, infants are born helpless, and unless adults among us feed them, shelter them, and keep them warm, they can't survive. Perhaps our species developed by selecting those among us who would take care of our young and others who are sick or helpless. In the 1600s, 1700s, and even 1800s, medicine so continued. It was practiced in monastaries and among nursing orders that provided succor, help, and comfort to those who were ill and in need.

Changing Role of Medicine

About 150 years ago, medicine began to change, and a new perspective was added. For the first time we began to look at *causes* of disease. Until then, medicine was descriptive—Laennec's "hard liver" was related to what we now know as cirrhosis; Bright described the small hard kidneys of nephritis, Bright's diease; and Addison described the abnormal adrenal glands of Addison's disease. If we want to date the change, we may look to Pasteur. With Pasteur, we began to find agents that caused disease, and this began a revolution in medicine. To emphasize how recently this actually took place, I would point out that 1982 was the 100th anniversary of the discovery of the tubercle bacillus. Diptheria, scarlet fever, dysentery, malaria, and other discoveries followed shortly thereafter.

This change in perspective carried with it the possibility that knowledge could be used to prevent disease. Within the next six decades, the public advocates of the time—they weren't necessarily scientists—pointed out to sometimes-reluctant state legislatures, concerned then as now with budget problems, that if they had sewer systems, pasteurization, and clean water, there would be a marked improvement in the public's health. Success followed, primarily in the first half of this century, and we continue to benefit.

Post-World War II Perspective

Doctors entered the post-World War II period full of optimism. We had seen that if you could discover the causes of disease, the maladies themselves could often be prevented. But we now found outselves in a totally different world. We began to identify noninfectious rather than infectious causes of disease. We identified vinyl chloride as a cause of hemangiosarcoma of the liver and asbestos as a cause of mesothelioma. We recognized that people who smoked cigarets often suffered emphysema and lung cancer. We found cancer associated with 4-aminobiphenal and with mustard gas. There was toxicity seen with lead and neoplasms with nickel smelting and arsenic. Scientists were very pleased with themselves for identifying these associations; it hadn't been easy, because most of the time we depended on epidemiological studies, which are relatively insensitive.

Unlike what happened in the 1920s, we were not greeted with universal acclaim. Rather, there was reluctance among some to receive such news. "Are you sure? Are you certain?" "How do you really know that Agent B kills people by causing bladder cancer? After all, Agent A is only slightly different, and it causes no cancer." "How can you be sure?" Recently, I lectured at Johns Hopkins and we discussed asbestosis, and what happened to asbestos workers. A physician rose and said, "I'm so and so from the Department of Radiology. What you are saying is not the case. The reason these people have asbestosis, scarred lungs, is that when they go home from work on a Friday night, they get drunk and vomit. This is aspirated. That's what gives them the scarring!"

Summary

We are now finding the causes of disease. This has particular importance because if one knows the cause of a disease, it is potentially preventable. This has led to an extraordinary paradox. Our generation is the first in human history to begin to learn the causes of cancer (we now know the causes of about 35 or 40% of all cancer). Yet, cancer is not being prevented. The information that we gain is often not utilized. It is clear that science is necessary, but not sufficient. Our generation has the potential to prevent preventable disease. With that opportunity comes responsibility. In a sense, health advocates have as great a responsibility as anyone to help translate scientific information into improved public health.

New Approaches to the Resolution of Professional Liability Problems

Don Harper Mills

Introduction

The medical malpractice insurance phenomenon is cyclical. California faced a crisis of hospital liability insurance costs as long ago as 1954, when all commercial carriers left the hospital field because of the worsening claims situation. Twenty years later, in 1974, we faced another crisis of both availability and dollars, affecting hospitals and physicians in many states. The medical and social response was so remarkable that many state legislatures were induced to pass a laundry list of reform legislation in an attempt to control the problem. For the first time we saw the development of insurance companies owned and operated by physicians. If these companies can ensure economic stability in hard times, the availability of insurance during the next cycle may not present a real problem.

Liability Today

Now in 1984, just ten years after the last crisis, we are facing a few cycle. The frequency and cost of the average claim are increasing, posing the threat of a severe economic burden on those who pay the premiums for professional liability insurance. It seems that although some of the tort reforms of the 1970s have proved meritorious, they are apparently providing an insufficient impact on malpractice litigation to hold down the cost of claims.

The effect of this new cycle in the late 1980s will be substantial. Until now we have been able to endure the increased premium costs by increasing fees to patients, but the advent of prospective payment for medical care will constitute a major roadblock. An economic crisis is sure to become a reality for physicians and hospitals. Such an outcome is inevitable if we continue the anomalous method of funding this type of litigation. The economic risk of medical malpractice litigation is a great social problem, yet it it funded through a single source—the health care participants (physicians and hospitals). This is hardly a case of spreading the risk.

Fault vs Nonfault Liability

A growing number of people want to solve this problem by changing the system of compensation from fault to nonfault liability. Some people believe, incorrectly, that we are already paying for most adverse outcomes suffered by patients; therefore, why not proceed into a nonfault system to avoid noneconomic recoveries? In 1976–1977, the California Medical Association and the California Hospital Association sponsored a study to determine the frequency with which a nonfault compensation system would be triggered, and how much it would cost. This study was a massive retrospective medical audit involving 23 hospitals and almost 21,000 patient charts. The study found that 4.65 percent of patients hospitalized in acute-care institutions in California suffered some form of adverse outcome of health-care management in the target year of 1974. Since there were three million patients hospitalized that year, this produced about 140,000 patients who would trigger the nonfault system. Actuaries then determined that the funding of such a system would cost a minimun of $800 million a year in California alone. This represents several times more than physicians and hospitals were paying in liability insurance premiums. Inflation since 1977 would probably now push the minimum figure above one billion dollars.

It is clearly not economically feasible to switch from a fault system to a nonfault system if the source of funding is going to remain the members of the health-care team. Therefore, regardless of the method of compensation for adverse medical outcome, if we have widespread prospective payment systems in place in the next decade, an economic crisis for the health-care industry will be too great to tolerate.

Funding Changes

Changing the system of funding will require tax support or a modification of the health-care insurance industry to combine hospitalization, major medical, and disability insurance with adverse medical outcome insurance. Although the potential burden of the cost of medical malpractice litigation may prove too great for the health-care industry, it would become only a small part of the entire health-care insurance market. Once the method of funding is altered, changing the concept of compensation from fault to nonfault becomes economically feasible, provided duplication of existing benefits can be eliminated.

To ask society to help us in our dilemma requires that we go forward with clean hands. Although our peer-review process is second to none among the professions, it can still be improved immensely. This is also true of the procedure for approving credentials, as well as for all quality assurance functions presently carried out at the hospital level. Those who feel that the malpractice litigation phenomenon helps to clear our house should be assured that we do not have to be induced by the fear of litigation to run a tight ship.

Summary

Managing adverse medical outcomes will play a significant part in the ability of hospitals (and perhaps physicians in the future) to survive the upcoming prospective payment system. If a patient develops a complication following a surgical procedure, the hospital will recover substantially less on the dollar for the treatment of that complication. Hopefully, we will react by focusing on identification and prevention of complications rather than on avoidance of high-risk procedures. As we will see, therefore, the method of health-care payment will become intricately entwined with the cost of managing adverse medical outcomes and malpractice litigation.

Advocacy for the Mentally Ill

Herbert Pardes

Introduction

It is typical to lament the difficulties advocating for mental health. Thirty-nine million people in the United States suffer from mental illness. Mental illness contributes to billions of dollars of waste in terms of people who are ill, the services required, and the drain on national productivity.

I would like to address the difficulties of advocating for mental health, the aspects of advocacy that involve the substance of mental health itself, the patient and the patient's relatives, the mental health provider, and the policymaker. Fortunately, we are progressing with regard to eliminating the stigma of mental illness. According to broad social surveys, more people are now willing to undergo mental health treatment than in past times; moreover, they come from a wider range of the socioeconomic spectrum. This, I believe, has been a consequence of the advocacy of mental health groups, as well as the media, which increasingly presents mental health and mental health treatment as a routine part of American life. Further, the Community Mental Health movement which created mental health programs throughout the country, brought people to understand that a mental health center in their town was no more unusual than a health-care clinic or a social services program.

The Subject—Mental Health

Advocacy for the mentally ill, and for the fight against psychiatric illness, is challenging for many reasons. Society frequently questions the morality and/or the personal strength of individuals who have psychiatric illnesses. Psychiatric illness is thus often seen as the result of some internal personal or family defect, rather than as the result of an external agent. If you catch a cold or develop pneumonia, most people do not stigmatize you. However, there does in fact seem to be a formidable stigma with regard to psychiatric illness, one that people seem to succumb to much too readily. I believe that many policymakers, like most people, assume that they will never develop a psychiatric illness, whereas they readily understand that they might someday develop heart disease or cancer. They thus conclude that it is better to put money to work against diseases that may well be personal dangers to them.

However, there have been other diseases and problems that have been the focus of stigma, and for which greater advocacy and greater resources have been successfully secured. In the past one whispered that someone had cancer. The mentally retarded were also, and to some extent still are, afforded less than the best support facilities. But advocates have worked hard for research on cancer and mental retardation, and these diseases have started to receive additional support. It is time for us in the mental health care field to look at how we present ourselves and to make our argument more effectively.

Potential Advocates

There is a problem regarding potential advocates. Often, patients are simply not able to advocate. When both you and society question your personal self-worth, how can you begin to take on the world. We have had unfortunate examples, such as a vice presidential candidate being put out of the running by virtue of his history of psychiatric illness. One might have commented at the time that depression is an eminently treatable disorder. We are able to effectively treat 85 to 90 percent of people with depression, and this vice presidential candidate has done well as a Senator. There are other examples of people feeling that their jobs were threatened, and people being adversely judged, by virtue of a history of psychiatric ilness. This again could prevent them from becoming open advocates. People working in corporations while in psychiatric treatment often have not claimed insurance benefits because of their fear that the company will know they have sought treatment. More recently, many have been willing to speak forcefully about mental health and for psychiatric treatment. Prominent people have been courageous in making public efforts: Joshua Logan, Dick Cavett, Jennifer Jones Simon, Burt Yancey, Rosemary Clooney, and Burt Lancaster have been articulate in speaking up for those afflicted with psychiatric illness. Paul Newman and Joanne Woodward have been extremely active after their child's catastrophe with drug abuse.

Very recently, Jack Hinckley, John Hinckley's father, converted what was a personal, family, and national tragedy into socially useful and adaptive behavior by developing the American Mental Health Fund for Research in Mental Health. Increasing numbers of people with varying degrees of national recognition have become active in this area. Also, it is noteworthy that the relatives and families of many patients have joined together to form organizations like the Alliance for Mental Illness, in New York State, and the National Alliance for the Mentally Ill. These people are working with the Mental Health Association, as well as with other groups, to create an effective citizen advocacy in behalf of the mentally ill.

Policymakers

Politicians are in many instances the most maligned of professionals. I think it is obvious that people in nearly any field seek certain goals. Businessmen seek profits, movie stars seek public recognition—so it is not surprising that politicians seek votes. But, politicians are also often concerned with trying to do good, and can listen to logic. There have been outstanding examples of politicians who have been helpful. Senator Mark Hatfield was responsible for getting six and a half million dollars into the 1982 National Institute of Mental Health (NIMH) budget. Silvio Conte was responsible in 1979 for a major boost in NIMH research and also for assuring that NIMH staff would receive pay raises from independent financial sources, rather than by taking money away from programs in mental health. I can report that there has been substantial support for mental health from Governor Mario Cuomo and many local Assemblypersons and Senators. Interestingly enough, in New York mental health is part of the large political picture, and I believe that most politicians can be made to listen if you can get them the facts.

In 1981, when the Reagan administration entered Washington, there was a threat to the entire National Institute of Mental Health program. We were told by the Office of Management and Budget (OMB) that their preliminary suggestion was to close all research programs, and training programs while service programs would be block granted (i.e., instead of a national competitive process directed by NIMH). We faced the desperate situation of having all of our grants terminated for 18 months. In other words, if a grant happened to come up during that 18-month period, funding would be terminated. We reversed that decision with an effective advocacy within the Department of Health and Human Services. We explained the facts and some of the advances with regard to mental illness. We took note of the recent recognition of mental health research in both the national and international health arenas. The result was that despite what might have been seen as a very challenging administration, the budget of NIMH for research has now reached $177 million. As an illustration of what happened in the past 16 years, it is noteworthy that from 1969 to 1977, the Mental Health Research budget actually went from 95 million to about 90 million. From 1977 to 1984, the budget went from 90 million to 177 million. In the first eight years, there was somewhat of a loss, while in the second eight years, there was almost a doubling. I submit that much of this is a result of better advocacy.

New Strategies

What do I mean by a new advocacy for mental health? Let me suggest some ideas:

1. We must not simply talk about patient suffering without discussing potential accomplishments, economically intelligent strategies, the excitement of our research, and the logic of our programs. I do not minimize the amount of suffering the patients and their relatives feel. But in the present atmosphere, it is not enough. Suffering is seen in many areas of the society and is a result of many different kinds of illness.
2. The psychiatric field is one of the most exciting health fields today as a result of the dramatic surge in knowledge from our research programs. In the last several years the psychiatric field has received two Lasker Awards and one Nobel Prize for work on the differential functioning of the two brain hemispheres, the PET scanner, and on the biology underlying behavioral functions.
3. The field is becoming increasingly differentiated and refined. No longer is a specific treatment whatever a psychiatrist can offer. Diagnosis is becoming refined by the use of operational definitions, such as those utilized in DSM-III.

Treatment is becoming specific for various conditions. We are increasingly using combined treatments. Data regarding the epidemiology of psychiatric disease is becoming more specific. We have made progress in disease areas, such as anxiety disorders, schizophrenia, and hyperactive disorders, but there is a tremendous way to go even as we make progress.

4. Mental health makes financial sense. In looking at off-set research, (i.e., the effects on utilization of health services when mental health services are offered), it has been demonstrated that the incorporation of mental health into general health services reduces by 20 percent the amount of general health services utilized. Mumford and Schlesinger (1) have shown that an appropriate incorporation of behavioral and psychiatric intervention in the general treatment of people with certain medical disorders, such as asthma, and other respiratory conditions or hypertension, also reduces the amount of general health services needed. Mumford and Schlesinger also pointed out that an appropriate incorporation of behavioral treatments when dealing with preoperative patients or those who have coronary disease, decreases their symptoms, as well as the length of their hospital stay.

Levitan et al. (2) demonstrated in an orthopedic setting, that when psychiatric service was part of the medical care, hospitalizations were reduced, patients were treated more comprehensively, and went home rather than to institutions more frequently (these patients were all over 65). The financial saving was approximately $190,000 for an investment in the psychiatric service of $10,000. Thus, a 19-to-1 savings occurred with more comprehensive care and patients going home earlier and undergoing less institutionalization. These all illustrate the logic of mental health services support. Research has shown that the introduction of lithium has saved $6 billion in the US during the last fifteen years. That is more than all the money ever spent on mental health research in this country. It is just one example of one development and an enormous saving.

5. Our treatments are being shown to be increasingly effective:

   (a) Psychopharmacological treatments are as well proven as any other treatment.

   (b) Psychotherapy is generally effective and we are now working to find out which psychotherapies are effective for which conditions.

   (c) Programs of psychosocial intervention, such as a Community Support Program, have been shown to reduce hospitalizations. One of the rewarding developments in psychiatry and mental health has been the integration of psychological, biological, and social interventions. We find that drugs, community support principles, and psychological work with the family all tend to help make the schizophrenic patient coming out of hospital less likely to require readmission.

6. Programs such as those providing transitional care have brought patients into work situations, and have resulted in less returning to hospital for additional treatment. The logic is that patients who rely upon support and resources from the community become workers themselves able to produce new resources. In addition, they feel a sense of pride in themselves by evolving into contributory members of society.

To repeat, I have said that this field is more effective, more refined, and more differentiated; its research is as vital as that of any field in health care, and the support for mental health programs makes good economic as well as health-care sense. I suggest that it is these arguments that should be used in talking to policymakers, the general citizenry, and the media.

Concluding Principles

There are a few other issues that deserve mention involving the collective group of mental health supporters. We must unite our groups—professional and citizen must work together. Professionals are often seen as self-serving when they advocate for mental health programs. However, providers have often been the only individuals who could advocate for patients. They are thus placed in the awkward situation of being damned if they do, and naive if they do not advocate. As a complement to organizing these groups, I suggest that all of us in mental health must be concerned about the entirety of mental health. I make it a point not to talk simply research to researchers and services to service people. One has to see that these programs help each other. It is the recognition of the enormous service need in mental health that justifies the support for research in mental health. At the same time, it is the excitement in research that helped us turn around what was in 1981 a dangerous perspective held by some of the new administration OMB people. We used the halo effects of the excitement of research and new refinements of treatment to help rescue the field.

It has been said that the brain is the last major frontier in health research. One of the most exciting aspects of this is the actual weaving together of higher intellectual functions of why we feel, how we think, and what we understand of its underlying biology. Kandel (3) has been one of the leaders in showing us that such issues can be researched. Dr. Kandel's and like research portends major advances, along with developments such as PET scans enabling us to look at brain activity, transplant work suggesting that brain damage may be reversed, receptor work showing that we can actually see which kinds of substances work in what parts of the brain and determine where people may be suffering deficiencies. We have genetics to show us where a diseased gene may be located and what its effects are.

There has never been a more exciting period in the field of mental health and brain research.

When one becomes concerned with whether we have made progress, it is important to look at the larger sweep of history. In the early 1940s, there were no antidepressants or tranquilizers, very little in the way of psychotherapy, and no lithium. For the most part, there existed a large number of remote institutions to which patients were sent. Practitioners did not know much of the nature of diseases and, as I said earlier, they often dispensed whatever treatment they knew, rather than attempting to match treatment specifically to the disorder.

Mental health practitioners can be more effective advocates because we not only have a stronger story to tell, but because we appreciate the importance of educating the public about the nature of mental illness. We can communicate the pain of a child, an adult, or a family experiencing the trauma of psychiatric illness. We must tell this story to the community, but let us also tell them that we have a way to address this kind of illness, that we have the promise of understanding, and that an entire society will be better mentally, socially, and economically because of the investment we make in mental health programs today.

References

1. Mumford, E., Schlesinger, H. J., Glass, G. V., Patrick, C., and Cuerdon, T. (1984) A new look at evidence about reduced cost of medical utilization following mental health treatment. *Amer. J. Psychiat.* **141,** 1145–1158.
2. Levitan, S. J. and Kornfeld, D. S. (1981) Clinical and cost benefits of liaison psychiatry. *Amer. J. Psychiat.* **138,** 6.
3. Kandel, E. R. (1979) Psychotherapy and the single synapse. *N. Eng. J. Med.* **301,** 1028–1037.

Communication Networks

Advocacy in Health Care

Suzanne Rauffenbart

Introduction

This article will focus on communication between the hospital and the patient, the doctor and the patient, and the media and the patient. I will discuss the role of communications in the health-care delivery system, exploring the opportunities that can be developed and implemented to improve communications and to make the system more effective.

The Role of Communication Between Hospital and Patient

From my observations around the country, there is frequently no systematic approach of communication to patients. In one institution I visited recently, there were three sets of signs instructing the patient where to go or what to do at the outpatient registration area—one on how to fill out complex forms at the registration desk, another telling the patient to see the financial counselors before seeing the doctor, and a third from the outpatient pharmacy explaining its hours of operation and billing procedures. The area looked like a mini-US

113

Route 1, where all the billboards compete for the driver's eye. All of the signs in the area were written to reflect what management needed to tell the patient, not what the patient needed to know.

Can you create a difference in patient attitude by changing the way you communicate to patients? When I was at Mount Sinai, we decided to find out by doing a comparative analysis of what I call a traditional means of hospital communication to patients versus a more patient-sensitive means of communication.

The hospital Admissions Office was chosen as the setting for the study. As in any Admissions Office, patients are filled with anxiety and tension: not knowing what to expect during the admitting procedures, having to wait a long time, and fearing the unknown in their diagnosis.

Some 80 elective surgical patients were selected for the study. Half of the patients were observed and interviewed following the traditional means of communication, and the other half were observed and interviewed following the more sensitive communication approach.

Basically, three major means of communication were put into place:

1. A clearly delineated sign welcoming patients and explaining in minimal and friendly language the first steps they should take.
2. A special information brochure designed with the patient in mind by a former patient who was a "consumer advocate" in her professional life. The handout was written based on observations and discussions with several patients going through the admissions process and other hospital procedures.
3. A quick and easy-to-read letter to the patient responding to frequently raised questions, explaining admitting procedures, and, very importantly, their rationales, and assuring patients they have not been forgotten. The letter was updated periodically as new questions arose.

These three simple communications steps achieved the following:

1. Those patients who received the special communications expressed more positive attitudes toward the hospital and the hospital's admission process—they appreciated that their waiting time was made easier.
2. Those patients who received the special communications were more self-reliant and did not have to stop to ask questions of the busy admission clerks and guards. In fact, patient-initiated interactions with staff were reduced by nearly half, freeing more time for the admitting staff to perform their jobs. And the staff felt better too—not having to interrupt their work for repetitive questions.
3. The level of anxiety was decreased. A comparison of EKG readings for both groups showed that the patients who received the special communications had slower heart rates, suggesting a lower level of anxiety.

Admissions is only the beginning of the relationship that the hospital has with the patient. I once counted 20 different communications to the patient for a standard hospital stay—creating a variety of tones, messages, and images of the institution. What do patients need and not need to know? How should the hospital position itself in these communications?

Doctors' Communications to Patients

A few years ago, a small group of journalists and public relations specialists met to discuss the image of doctors and hospitals throughout the country. Among the works we explored was a study from a major teaching hospital of how people form their opinions about physicians.

The researchers found that the majority of people formed their opinions of physicians and hospitals based on their own or friends' and relatives' experiences with the physician or hospital. Only 15 percent formed their opinions of the physician or the hospital based on information from the media.

What was especially interesting was that physicians had a very different view of what their patients were thinking. The hospital physicians perceived the public image of the institution to be much more negative than the study results showed.

Recent studies have also shown that physicians have little idea of their impact on their patients. Take, for instance, a 20-minute office visit, in which sociologists observed less than one minute was spent by physicians informing patients about their medical problem or treatment. Patients, following their visits, were quite aware of this. The physicians, however, reported that they thought they spend an average of 10 to 15 minutes giving information.

How physicians perceive their relationship with their patients is becoming a fertile field for study. Medical students are now learning "bedside manner" in medical school courses such as Humanistic Medicine. In medical rounds that I have been on in recent times, I have noticed attending physicians discussing with the residents the dynamics of the physician–patient relationship and how to improve it.

How can that be achieved? What are the responsibilities of the physician and patient in establishing and maintaining that relationship?

The Media and the Patient

Over the last several years, a new line of communication has been established with the patient. The growing body of medical and scientific media has provided an unparalleled source of information to patients, bringing to their attention new methods of diagnosis and new forms of therapy. On the public policy front, there have been documentaries on the high cost of health care. In the financial pages of our newspapers, we read that the for-profit hospitals—and their outstanding growth potential—have aroused investors' interest. The economics of the health-care system, previously considered incomprehensible because it didn't fit the traditional demand/supply market model, is being discussed, heightening the public's attention to the complexities of the system, the highly regulated aspects of health-care delivery, and the difficult choices that need to be made in the public-policy arena.

What effect does this increasingly expanding source of information have on the patient? How has it changed the physician–patient relationship? Is it creating unrealistic expectations? Has it changed the system?

The press in Washington—with all of its power and influence on the legislative process—has come to be known as the fourth branch of government. Likewise, the emerging role of the press in medical care could soon join the patient, the doctor, and the hospital as the fourth branch of the health-care delivery system.

The Patient's Role in the System

Up to now, we have focused on the hospital's communication to the patient, the physician's communication to the patient, and the growing role of the media in communicating to the patient. There is another key communications component in the health-care system that will also be discussed—the patients' communication role in the health care system. What are the responsibilities of the patient? How active a role should patients play in their care? What can be accomplished? What risks do the patients run?

Richard Block, better know and most familiar to us as the R in H & R Block, has launched a personal campaign to encourage an activist role among patients. He himself did it and scored a major victory. Several years ago, he was diagnosed as having lung cancer and given a dim prognosis. He wasn't satisfied with the answer he got—he tried and tried again, pursuing other doctors, other therapies. Once he recovered and was considered "cured," he started his own campaign to encourage patient activism. He leaves his imprint wherever he goes:

> *"To obtain the maximum benefit from your treatment, become a partner with your physician. Understand everything he or she is doing for you and how and why it works. To do this you must ask questions. You physician wants you to understand but may not know what to explain."*

Question, Comments, and Answers

In addition to the published proceedings of the papers originally presented at the Advocacy in Health Care Conference, our seminar included an afternoon discussion session. During the discussions, both speaker–participants and the audience were encouraged to question the keynote speakers and seminar participants. The following selections represent excerpts from the dialog between attendees and invited speakers.

Comment: One principle that we should be advocating is for everybody in this country to have adequate health care insurance coverage. You can see that the outrageous care given to indigent people is partly a function of the fact that they lack such health care coverage. The effect of an advocacy for universal health care cannot be predicted, but certainly the principle must be emphasized. Now, if you are looking to find the money for universal coverage in the system, aside from some currently absorbed by the Defense Department that would be eminently suitable for this purpose, one place that might be looked at is health-care reimbursement. There

119

is a disparity between technology and procedure reimbursement versus patient–interaction reimbursement. In other words, if a family practitioner, pediatrician, or psychiatrist spends an hour with the patient, the reimbursement is very modest. But, if a specialized surgical operation is performed, the providers will receive perhaps 50 to 100 times as much for the same hour of care. And you could very easily, if you had control of the reimbursement system, reduce that inequality. By increasing some of the other services you could probably reduce the tendency to use technology excessively. As for the government running programs, I think the government sometimes becomes the only avenue by which we can address problems that other parts of society will not address. I believe that you have to pay attention to government's internal rigidities, which often make its functioning less than optimum.

Comment: With a $300 billion deficit, I think you ought to be reacting loudly and clearly out there. I think you ought to be incensed. You are dealing with matters of life and death. You are getting the short end of the stick. Your needs, or the needs that you represent, are not being met. So, I do not want to be sensitive while there is a $300 billion debt budget. I do not have to be sensitive about that. I think the sensitivity ought to be on the other side. In many hospitals, people die because they are unattended: it's that simple. Now if that is emotional, hell, maybe we need a little more emotion. But, I don't want to measure it. I don't want to be sweet while there's an allocation of resources going the wrong way. I think you ought to be angry. I think you ought to grind the bastards down. I don't think you ought to sit back or be nice about it.

Comment: I'd like to support the comments that it's time to get angry at a $300 billion budget deficit when our community needs are not being met. I think its commendable that people are sitting in the same room and talking to each other—because health cannot be isolated from the economy. *Advocacy is a political process.* Hospitals necessarily become political animals. But how does a community advocate for itself?

Comment: Regardless of where you are politically, are there ways to blend program and economics? I would submit to you that while I'd be happy to fight and march behind your banner to get what you'd like, it's going to be politically difficult. We have to think not whether there's a mild or moderate or sensitive way, but whether there is an effective way of making health policy in the country.

Comment: Our whole health system at this point—speaking as a child health advocate—is in serious disarray. We have a system in which, depending on what kind of job you have, depending in what region of the country you live, depending on whether you happen to be poor, you're either going to or not going to have access to a doctor. There are some fundamental truths that seem to come out again and again and again. One is that, if you're poor, you're not particularly healthy, no matter what measures are taken. That if you're poor, you're at greater health risk all the way down the line. There's just so much education you can give poor people about how to be healthier. They have very few choices. I believe we have rationed health care in this country for as long as there's been any organized health system delivering

health care, and as long as health care
has been viewed as a benefit to be got-
ten. And I agree completely with the
speaker who said, "until people turn
around and say you cannot run a health
system this way nothing will change." It
cannot be a mere stroke of fate whether
you live in South Carolina or New York
that your child's going to have a
Medicaid card or not, or whether you
happen to have a job with the Rand Corporation or as custodian for McDonalds,
that your child is going to get immunized—that's what's got to stop. And I
don't see the anger in this room that
should be there from a group of people
who've had to work with sick people,
with poor people, with old people.

Comment: Until we come to grips with the
semminal fact that the health care system now is not being driven either by
need or opportunity, but by dollars, we
are not being realistic. We heard a lot
about the present administration—and
I wouldn't disagree with what we
heard—coming out very badly in terms
of priorities for allocation to health care
and for prevention. But, we also heard a
Democratic appointee to the Commission of Health, if you listened carefully,
driving the system by the dollar. We
should be angry that, at a time when opportunities have never been greater to
intervene, both in terms of prevention
as well as effective care management,
we are being forced to focus on saving
the buck. We're going to pay for that
down the pike when the public becomes
outraged at the increasing gap between
what can be done for those who can afford it and what is being done for most
of the people. There is no substitute for

placing a higher priority on the allocation of resources for health if we want better disease prevention and care. We lose sight of this because of the enormous pressure to cut back costs. It's true about all of us in the health field. We just never have the opportunity to sit back and say this is outrageous. In a larger sense, this reflects the public's views. I do believe in democracy and I think the public is receiving what it is asking for, so to speak. There just isn't enough public outrage at the lack of resources for health and perhaps this reflects the fundamental fact that the public doesn't consider health a priority. The vast majority of health professionals are committed to a healthy nation.

With respect to prevention, we have to admit that we really don't know how to intervene in the most effective way. It is not just a question that it is going to take 10 or 20 years to resolve. We don't know how to do it effectively. As an example, take smoking. The evidence is overwhelming that smoking is the single most important environmental contributor we can identify as a cause of cancer. Still we don't know how to effectively decrease smoking in this country—that is, how to change behavior.

Comment: I'd like to underline something said. In the District of Columbia—and I am in the administration of the Medicaid program—we have a $300 million budget for Medicaid and we have one of the largest Medicaid populations, for our size, as well as a very generous program. Last year I, too, heard all the cost-containment rhetoric that we are about to slash out all the optional services for Medicaid and then the advocacy groups came out of the woodwork. We had legal services for the elderly. We had the mental health advocacy group. And one of the other things that happened was we had one of the advocacy groups going to court and we got the judge to practically decree to empty Forest Haven of the mentally retarded, and then the money came forth. The mayor said no, we're not going to touch the optional services—we're not going to do this, that, or the other—and the money came out. There *is* money there to support programs, but there has to be more than one advocate, more than one person. There have to be several groups that cross subpopulations. One person can't do it, one group can't do it, but a number of groups all working toward the same end can make a difference.

Question: I'm interested in information. We can get it on the city and state level. Our problem is being able to get it on the federal level as a basis for comparison. If we don't have the information, we're not able to say that the statistics are biased, or compiled in negative ways for a community. We see skewed, false mortality rates and police department data. Only by our being able to make the comparison of city, federal, and state data can we then be effective. Is the federal government closing up more avenues of information so that it is no longer going to be available, so comparisons are not going to enable us to make meaningful comparisons? We need the federal data.

Comment: I think that's very important. When I was in Washington, we realized there were no usable health data, for example, on Hispanics. We immediately got to work to uncover the reasons. I think what you are saying is that we must ensure there are valid basic data available on the various ethnic groups.

Question: I wonder if I might ask Paul Rogers to comment on the interrelationship of science and advocacy. Science can be seen as the ultimate cost containment and the ultimate service. We no longer have TB sanitaria and we don't have iron lungs. Would you comment on the interrelationship of the research dollar and advocacy?

Mr. Rogers: Of course there is a very definite relationship between the research dollar and advocacy. One of our problems used to be where should we try to allocate the research dollar, where would it do the most good? People want an immediate response, they want to see something happen—the cost–benefit ratio—they want it established immediately. And that's one of the problems in this nation. This is true in research and prevention and we often don't see the results of prevention program for 20 years. Often people want us to do something in research—operate and fix it up right now. So, it' going to take a lot of advocacy explaining and understanding to begin to shift the emphasis on research where it may do us the most good in the long run and I think this is a role health advocacy can also play.

Question: Mr. Gotbaum, what do you see in the future for unionized health care workers, especially in light of the industry's advances in technology and what that has meant to jobs and the ability to absorb changes in the system.

Mr. Gotbaum: The best way of projecting the future is to see what's happening now. Unionization will continue and will grow. Union organizing, though, is where you have your manpower. Health services are growing. There's going to be more of an emphasis on organizing. The workforce, especially in the service area, is becoming more heavily women. Your most loyal members are blacks and Hispanics. The minorities are the most loyal members of the union. We don't have enough leadership in that area. We don't have enough staff in that area. We will. We're a little awkward, sometimes a little vulgar, but I think that in terms of our goals, they're good goals and they're meaningful ones, and that's about it.

Comment: I would like to make an observation with regard to a conflict that frequently arises when the issue of patient representation is discussed. The conflict between the patient representative's mission to assist the patient or to be an advocate to the patient and the role of the patient representative as an employee and part of the staff of the hospital. Cynics sometimes say that this conflict totally vitiates the concept of patient representation. I don't believe this is the case. Patient representatives are part of the institution, they cannot act independently. Their ability to achieve goals depends on working harmoniously and effectively with the entire institution, and if they don't work in this way they cannot achieve their objectives.

Frequently, we think of the patient representative in a potentially adversarial relationship, but this is not usually the case. An additional role that the patient representative plays is to step in when there is no problem, no complaint. The resources of the hospital are varied and complex and frequently new services and programs are introduced. Staff taking direct care of patients may not be aware of newly available services or may not be able to identify what services might benefit particular patients.

Question: Should a patient representative see all patients who come into the hospital whether or not they have a complaint?

Answer: In some hospitals, on admission patient representatives see or send special messages to every patient. Some institutions survey a random sample of patients on various nursing units. Questions are asked about services and in that way one hears about current problems and learns how patients feel about the hospital and about services. For the patient representative to see every patient, you would need a very large department. I think you need to take the squeaky wheel and use the problems presented to oil the system. One hopes that the entire system will get better in response to the specific patient needs revealed by this process. Many hospitals use a hotline that directly connects the patient to the patient representative. Often there is also patient representative referral information laminated onto the telephone and on the same page as the Patient's Bill of Rights in the patient's information booklet.

Question: What authority does a patient representative usually have in a hospital?

Answer: On paper, actually very little, but authority lies in knowledge of the formal and informal rules, knowledge of which staff people will be able and will want to help. On paper it's not authority, but an actual working relationship. There is a good deal of power in having an overview of the system and knowledge of the staff.

Question: Do patient representatives see the area of prevention as part of the role?

Answer: Yes! One of our key preventive roles is to change the system so that it becomes better for all the patients, using the squeaky wheel to oil the system, so that it is smoother and more effective in its responses. Many patient representatives are active in the health educational committees in their hospitals. At Mr. Sinai in New York City, for instance, the patient representatives have focus groups to discuss patients' problems and are in the midst of producing a videotape showing patients how to understand patients' rights and how to ask questions of their doctors. There are task forces within the hospital making plans for educational material that will be presented to patients.

Bibliographies

David Axelrod, M.D. Commissioner, New York State Department of Health since 1979. Previously served as Director, Division of Laboratories and Research, New York State Department of Health; President, Health Research, Inc.

Robert N. Butler, M.D. Brookdale Professor of Geriatrics and Adult Development. Chairman, the Gerald and May Ellen Ritter Department of Geriatrics and Adult Development, Mount Sinai School of Medicine, New York City. Founding Director of the National Institute on Aging of the National Institutes of Heath. Author and/or co-author of: "Human Aging" 1963; "Aging in Mental Health" (with Myrna I. Lewis) 1973; "Why Survive Being Old in America" (Pulitzer Prize 1976); "Love and Sex After Sixty" (with Myrna I. Lewis) 1975.

Victor Gotbaum Executive Director of District Council 37, American Federation of State, County and Municipal Employees AFL-CIO. He serves on the Board of Directors of the Regional Plan Association and the Citizens Committee for New York City. Vice-President of the New York City Central Labor Council, AFL-CIO.

David Steven Greer, M.D. Dean of Medicine, Professor of Community Health, Brown University School of Medicine, Providence, Rhode Island. A member of the Institute of Medicine of the National Academy of Science, Consultant to the Governor and board member of the Health Planning Council of Rhode Island. Founder of the first hospital for chronic diseases and rehabilitation in Southeastern Massachusetts and the first medically-oriented public supported housing project for physically disabled persons.

129

Myrna I. Lewis, C.S.W.   Faculty, Department of Community Medicine, Mount Sinai School of Medicine, New York City. Psychotherapist, social worker (ACSW), writer, gerontologist. Author and/or co-author of: "Aging in Mental Health" (with Robert N. Butler) 1973; "Love and Sex After Sixty" (with Robert N. Butler) 1975; "The History of Female Sexuality in the United States" in "Women's Sexual Development" (Edited by Martha Kirkpatric) 1980; "Conversation with a Geisha" (with Masado Osako) in press.

George Lythcott, M.D.   Edward Jenner Professor of International Health, University of Wisconsin-Madison School of Medicine, Madison, Wisconsin. Professor of Pediatrics. Director, Center of International Health Resources and Services. Former Administrator of the Health Services Administration (DHHS), Washington, D.C. and Assistant Surgeon General (United States Public Health Service), Washington, D.C. Dr. Lythcott has served as an alternate U.S. delegate to UNICEF and a member, U.S. delegation to WHO General Assembly.

Joan H. Marks, M.S.   Co-Director and founder, Health Advocacy Program, Sarah Lawrence College, Bronxville, New York, Director, Human Genetics Program, Sarah Lawrence College. Author of "The Genetic Connection: How To Protect Your Family Against Hereditary Disease" 1978.

Don Harper Mills, M.D., J.D.   Diplomate of the American Board of Law in Medicine; Clinical Professor of Pathology, University of Southern California School of Medicine; Attending Staff (Pathology) Los Angeles County General Hospital; Panel of Arbitrators American Arbitration Association; Project Director California Hospital Association Notification System. Executive Editor, *Trauma;* Editorial Boards of the *American Journal of Forensic Medicine and Pathology, Hospital Risk Management, Hospital Risk Control.*

H. Richard Nesson, M.D.   President, Brigham and Women's Hospital, Boston, Mass., First Medical Director of the Harvard Community Health Plan.

Herbert Pardes, M.D.   Director, New York State Psychiatric Institute, New York City; Chairman of the Department of Psychiatry at Columbia University. Director of the National Institute of Mental Health 1978–1983.

Suzanne Rauffenbart   Vice-President of Public Affairs, Memorial Sloan-Kettering Cancer Center, New York City. Previously served as Vice-President of Public Affairs at Mount Sinai Medical Center, New York City.

Ruth Ravich   Director, Patient Representative Department, Mount Sinai Medical Center, New York City. Co-Director and faculty member, Health Advocacy Program, Sarah Lawrence College, Bronxville, New York. Founder and first President of the Society of Patient Representatives of the American Hospital Association. Author or co-author of: "Informed Consent to Biomedical Research in Veteran's Administration Hospitals" (with Henry W. Reicken); "Patient Relations"; "Ombudsman Program Provides Feedback" (with Helen Rehr).

Paul G. Rogers, J.D.   Partner in the law firm of Hogan & Hartson, Washington, D.C. Served 24 years in the United States House of Representatives, 8 years as Chairman of the House Committee on Health and the Environment. Chairman, National Council on Patient Information and Education. Chairman, Advisory Council on Pre-paid Managed Care for Medicaid Recipients, a project of the Robert Wood Johnson Foundation.

Sara Rosenbaum, J.D.   Director, Child Health Division, Children's Defense Fund, Washington, D.C. Previously served as Staff Attorney for the National Health Law Program and for the Vermont Legal Aid Association. Recipient of the Health Care Financing Administration's Beneficiary Award for distinguished national service on behalf of medicaid beneficiaries.

Irving Selikoff, M.D.   Professor of Community Medicine; Director, Environmental Science Laboratory and Program Director, Environmental Health Science Research Center, Mount Sinai School of Medicine, New York City. Governor, New York Academy of Medicine; Member, National Cancer Advisory Board; Editor-in-Chief, *American Journal of Industrial Medicine + r* President, *Collegium Ramazzini*.

Vergil Nelson Slee, M.D., M.P.H.   President, Slee Associates (Quality Assurance Consultants), Ann Arbor, Michigan. President Emeritus, Commission on Professional & Hospital Activities. Writer and lecturer, as an authority on classifications of disease, helped develop the International Classification of Disease. Past-President of the Council on Clinical Classifications.

Index

Abel-Smith, Brian, 63
Advocacy, *see* Patient
  advocacy
Aging
  advocacy for, 61–65
  budget priorities, 62–64
  high health costs of, 65
Alliance for Mental Illness,
  103
Alzheimer's disease, 61
Ambulance drivers
  involvement in patient
    care, 80–81
American Association of
  Retired Persons, 73
American Hospital
  Association, 51, 52
  patient's Bill of Rights,
    53–54
American Society for the
  Prevention of Cruelty
  to Animals, 2
Asbestosis, 95

Black Americans
  infant mortality, 28–29
  older women in poverty,
    70
  prenatal care, 30–31
Bleich, J. David, 18

Blue Cross, 25
Bob and Ray, hospital radio
  skit, 16
Brain, research on, 108–109
Budgets
  aging priorities, 62–64
  Medicaid, 121–122
  mental health, 104–105

California Medical
  Association
  nonfault compensation
    study, 98–99
Cancer patients,
  warehousing of, 24
Cardiovascular disease, *see*
  Disease
Cavett, Dick, 103
Children
  as animals, 2
  effective advocacy for,
    47–49
  health risks among, 45–46
  poverty's effects, 41–45
  public health programs,
    cutback, 47
  research involving, 3
Children's Defense Fund,
  43–44, 49

133

China
  infant mortality, 28–29
  ombudsman in, 55
Clooney, Rosemary, 103
Communications
  doctor to patient, 113–114
  hospital and patient,
    111–113
  media and patient,
    114–115
Community Mental Health
  movement, 101
Confidentiality
  patient's Bill of Rights, 54
Conflict of interest
  patient advocates, 5
Consent, *see* Informed
  consent
Consumer
  health, education of,
    16–18
Coronaries, *see* Disease,
  cardiovascular
Cultural diversity
  Diffusion Strategies
    Report, 35–36
  minorities, 32–34

Death with dignity, 6–7
Delbeco and Nominal
    Group Technique, 89
Diagnostic Related Group,
    88–90
  patient representatives
    and, 58–59
Diffusion Strategies Report,
  35–36
Disease, *see also* Mental
  illness
  Addison's, 94
  Alzheimer's, 64

Bright's, 94
cardiovascular, 32–33, 36
database faults, 31
paradox in, 95
Doctors, *see* Physicians
DRGs, *see* Diagnostics
    Related Group
Dying, *see also* Saunders,
    Dr. Cicely
  as part of living, 22
  demographic changes in,
    21–22

Elderly, *see also* Aging
  health care's neglect of,
    15
  research involving, 3
Ethics, Robin Hood ethic,
    18
Ethnic health comparisons
  database faults, 31, 38

Feedback loops, *see* Quality
    (of care)
Food stamps and older
    women, 72
Forbes magazine, 63
Forms, consent, 5

Geriatrics and budget
    priorities, 63–64
Ginzberg, Eli, 15–16
Gray Panthers, 73
"Great Equation," 14
Grievance mechanisms, 54
Group advocacy, *see* Patient
    advocacy

H & R Block, 116

Handicapped
  patient representatives
    and, 58
Harlem
  health information
    diffusion, 37
  mortality rate, 15
Health care, 28–34, *see also*
    Quality (of care)
  costs, 17
  etiquette of, 11–12
  minorities, issues and
    strategies, 38–39
  organizing workers in,
    78–79
  physician overabundance,
    13
  rationalization of, 14
Henderson, Lawrence,
    11–12
High Blood Pressure
    Education Program, 35
Hill-Burton program, 43
Hinckley, Jack, 103
Hispanics
  health data, 122–123
  infant mortality, data
    lacking, 28–29, 31
  older women in poverty,
    70
Hospice, *see also* Patient
    advocacy
  described, 20
  expanding role of, 21–23
  first American, 19
  professionals in, 24–26
  services provided, 20–21
  volunteers in, 25–26
Hospitals, *see also*
    Communications;
    Medical malpractice,
    liability insurance;

Patient advocacy;
    Patient representatives;
    Quality (of care)
  accreditation surveys, 91
  Board's duties, 85, 90–91
  Bob and Ray radio skit,
    16
  high-quality described, 93
  investor-owned
    proprietary, 9–10
  patient representative
    program, 51
  poor people, case
    histories, 41–43
  unionizing workers in,
    77–82
Hypertension, minority
    forum on, 35

Illness, older men and
    women compared,
    68–69
Indians, American
  health information
    diffusion, 37
  infant mortality, 28–29
  Navajo hip disease, 31
Industrial Revolution, 63
Infant mortality
  American Indians, 28–29
  black Americans, 28–29
  Chinese/Japanese, 29
  Harlem, 15
  Hispanics, 28–29
  poor children, 44–45
  white Americans, 29
Inflation, increase in, 65
Informed consent
  patients, 5–6
  patient's Bill of Rights, 54
Inglefinger, Dr. Franz, 12

Institutional Review Boards
  patient representatives
    on, 57
Intervention, medical
  success of, 12

Japanese, infant mortality,
  28–29
Juran, J. M., 88–89

Kandel, E. R., 108
Kubler-Ross, E., *On Death
  and Dying,* 22

Lancaster, Burt, 103
Lasker, Mary, 75
Legal advocate
  patient representative
    contrasted, 56
Levitan, S. J., 106
Liability insurance, *see*
  Medical malpractice
Life expectancy
  gains in, 62
  whites/nonwhites, 29
Logan, Joshua, 103
Longevity Revolution, 62–65
  women, 67–68

Mahoney, Florence, 75
Mandex report, 31, 35
  Navajo hip disease, 33–34
Maternal mortality, whites/
  nonwhites, 30
Media, patients affected by,
  114–115
Medicaid, 25, *see also*
  Medicare

budget, 121–122
  older women in, 71
Medical care, health
  equated, 14
Medical Ethics Committees,
  58
Medical malpractice
  liability insurance
    cyclical nature of, 97–98
    fault vs nonfault
      funding, 98–100
  patient grievance
    mechanisms to avoid, 2
Medicare
  hospice, 25
  older women and, 70
  political fights over, 71–73
  poor children, 46, 47
Medicine
  changing role of, 93–94
  defined, 11
  origins, 93
  post-WW II perspective,
    94–95
Menage, Gilles, 11
Mental illness
  advocacy for, 102–103,
    105–107
  new strategies, 105–107
  politics of, 104–105
  stigma of, 102
Mental patients, research
  involving, 3
Minorities, *see also* Black
    Americans; Chinese;
    Hispanics; Indians,
    American; Japanese
  cultural diversity among,
    32–33
  diseases among, 32–33
  health care issues and
    strategies, 38–39

health information
diffusion, 34–39
loyalty to union, 124
"Miranda" warnings:
patients' rights, 5
Mortality, *see* Infant
mortality; Maternal
mortality; Minorities
Mumford, E., 106

National Alliance for the
Mentally Ill, 103
National Commission for
the Protection of
Human Subjects of
Biomedical and
Behavioral Research, 3
National Counsel of Senior
Citizens, 73
National Heart, Lung and
Blood Institute, 35–36
National Hospice
Organization, 20, *see
also* Hospice
National Institute of Aging,
64
National Institute of Health,
budget, 64
National Society of Patient
Representatives, 51
Navajo Indians
congenital hip dislocation,
34
*New England Journal of
Medicine*, 12
New York State Health
Code
patient's Bill of Rights,
53–54
New York, physicians in, 13
Newman, Paul, 103

Nominal Group Technique,
89
Nurses
as angry workers, 79
organization of, 81
Nursing homes, older
women in, 69
Nursing Outlook, 51

O'Connor, Justice Sandra
Day, 74
Old age, *see* Aging; Elderly
Older women, *see also*
Aging; Elderly
class factors of poverty,
71
financing of health care
for, 70
fortunes in hands of, 75
illness and disability
patterns of, 68
in nursing homes, 69
in politics, 74
longevity of, 67–68
men compared, 67–68
organic brain disease, 69
political issues in, 71–74
reasons for, 68–69
verbal abuse of, 75–76
widowhood, average age
of, 71
Older Women's League, 73
Ombudsman, patient
representatives based
on, 55
OPEC, 63
Osteoporosis, 64

Paraprofessionals
(hospitals)

involvement in patient
    care, 80–81
unionizing of, 79–80
Pareto principle, 89, 92
Pasteur, Louis, 94
Patient advocacy
    advocacy defined, 23
    as lost priority, 2–3
    conflict of interest, 5
    dying patients, 6
    group advocacy, 23–24
    hospice advocacy, 23
    poor children, 43–44,
        47–49
    problems influencing role
        of, 10
    professional and
        volunteer, 24–26
    training needed, 4
Patient representatives
    authority of, 125–126
    causes leading to, 52–53
    conflict of interest, 57,
        124–125
    development of field,
        55–59
    patient's Bill of Rights,
        53–54
    role of, 55–59, 126

Patients, *see also*
        Communications;
        Patients' rights
    media's effect upon,
        114–115
    physician's impact on,
        114
    role in system, 116
Patient's Bill of Rights,
        53–54, 125
Patients' rights

Johnny Carson's parody
    of, 4
"Miranda" warnings, 5
movement, 3–4
residential health care
    facilities, 16
written code of, 16
PET scanner, 105, 108
Physicians
    as facilitator, 12
    "bedside manner," 114
    communications with
        patient, 113–114
    impact on patient, 114
    oversupply of, 13–14
    overutilization of
        technology, 21–22
    patient representatives,
        52–53
    redefining role of, 14–15
    specialization's
        drawbacks, 53
Political office, older women
    in, 74
Poor people, *see also*
        Children
    health care's neglect of,
        15
    hospital problem, 41–43
Poverty
    black older women in, 70
    class factors of, 71
    "new poverty," 72
    rate of, 72
Pratt, Louis, 15
Pregnant adolescents, 14
    health care's neglect of,
        15
Prenatal care
    black/white comparisons,
        30–31
    poor children, 45

President's Commission for the Study of Ethical Problems, 46
President's Commission on Medical Malpractice, 52
Press, *see* Media
Prisoners, research involving, 3
Privacy, patient's Bill of Rights, 54
Professionals, hospice advocacy, 24–26

Quality (of care), *see also* Hospitals
accountability for, 91
control defined, 87–88
defined, 83–84
feedback loops, 85–88
function defined, 85
improvement defined, 86
laws regarding, 84, 91–92
management audit, 90–91
management's task, 92–93
protocols in, 88–90
responsibility for, 81
Questionnaires, patient representatives and, 58

Reagan administration
health care cutbacks, 120
Medicare/Medicaid cuts, 71–72
mental health budget cuts, 104–105, 108
women voter's for, 73–74
Reinhardt, Uwe, 11
Research, patient advocacy to monitor, 3
Roberts, Lillian, 79

Robin Hood ethic, 18
Roemer, Milton, 17
Rogers, Everett, 33

Sarah Lawrence College health advocacy program, 52
Saunders, Dr. Cicely, 19, 23, 26
Schlesinger, H. J., 106
Science, advocacy's interrelationship with, 123
Select Panel for the Promotion of Child Health, 46
Sign-language interpreter, as patient representative, 58
Simon, Jennifer Jones, 103
Smoking, 121
Social Security, crisis in, 73–74
State legislators, women as, 74
Supplementary Security Income, older women, 72
Sweden, ombudsmen's role, 55

Tax Equity and Fiscal Responsibility act, 25
Technology
cut back opposed, 65
overutilization of, 21–22, 24
specialization's drawbacks, 53
The Hospital Medical Staff, 51

Triage, 62

Unions
    health-care workers,
        78–79
    minorities' loyalty to, 124

Volunteers, hospice
    advocacy, 25–26
Voting, older women, 74

Washington Heights, health
    information diffusion,
    37
White Americans,

infant mortality, 29
life expectancy, 29
maternal mortality, 30
older women in poverty,
    70
prenatal care, 30–31
White House Conference
    "Responding to the
    Health Care
    Consumer," 52
Widowhood, average age
    of, 71
Wildavsky, Aaron, 14
Women, *see* Older women
Woodward, Joanne, 103

Yancey, Burt, 103